The aim of this book is to develop a unified approach to nonlinear science which does justice to its multiple facets and to the diversity and richness of the concepts and tools developed in this field over the years.

Nonlinear science emerged in its present form following a series of closely related and decisive analytic, numerical and experimental developments that took place over the past three decades. It appeals to an extremely large variety of subject areas, but, at the same time, introduces into science a new way of thinking based on a subtle interplay between qualitative and quantitative techniques, topological and metric considerations and deterministic and statistical views. Special effort has been made throughout the book to illustrate the development of the subject by physical examples and prototypical experiments, and the mathematical techniques by reference to simple models. Each chapter concludes with a set of problems.

This book will be of great value to graduate students in physics, applied mathematics, chemistry, engineering and biology taking courses in nonlinear science and its applications, as well as to researchers and teachers involved in one way or another in this field.

Introduction to nonlinear science

Introduction to nonlinear science

G. Nicolis

University of Brussels

CAMBRIDGE
UNIVERSITY PRESS

CAMBRIDGE UNIVERSITY PRESS
Cambridge, New York, Melbourne, Madrid, Cape Town, Singapore, São Paulo

Cambridge University Press
The Edinburgh Building, Cambridge CB2 2RU, UK

Published in the United States of America by Cambridge University Press, New York

www.cambridge.org
Information on this title: www.cambridge.org/9780521462280

First published 1995
Reprinted 1999

A catalogue record for this publication is available from the British Library

Library of Congress Cataloguing in Publication data

Nicolis, G., 1939–
Introduction to nonlinear science/G. Nicolis
    p. 13 cm.
ISBN 0-521-46228-2. – ISBN 0-521-46782-9 (pbk.)
1. Science.   2. Nonlinear theories.   I. Title.
Q172.N53   1995
500–dc20   94-15302   CIP

ISBN-13 978-0-521-46228-0 hardback
ISBN-10 0-521-46228-2 hardback

ISBN-13 978-0-521-46782-7 paperback
ISBN-10 0-521-46782-9 paperback

Transferred to digital printing 2005

To Katy,
a dependable attractor in a turbulent world

Contents

Preface

Nonlinear science emerged in its present form following a series of decisive analytic, numerical and experimental developments that took place in close interaction in the last three decades. Its aim is to provide the concepts and the techniques necessary for a unified description of the particular, yet quite large, class of phenomena whereby simple deterministic systems give rise to complex behavior associated with the appearance of unexpected spatial structures or evolutionary events. Such systems are encountered in a great number of disciplines notably in classical mechanics, statistical physics, fluid dynamics, chemistry, optics, atomic and molecular physics, environmental sciences, engineering sciences or biology, in the context of both fundamental and applied investigations.

While the concern for unification is central in every attempt of man to explain the natural world, the particular approach followed by nonlinear science in the pursuit of this goal is characterized by a great originality that differentiates it from other disciplines belonging to the traditional realm of physical sciences. Nonlinear science introduces a new way of thinking based on a subtle interplay between qualitative and quantitative techniques, between topological, geometric and metric considerations, between deterministic and statistical aspects. It uses an extremely large variety of methods from very diverse disciplines, but through the process of continual switching between different views of the same reality these methods are cross-fertilized and blended into a unique combination that gives them a marked added value. Most important of all, nonlinear science helps to identify the appropriate level of description in which unification and universality can be expected. The fundamental laws of microscopic physics such as Newton's equations or Schrödinger's equation, or of macroscopic physics such as the Navier–Stokes equations or the law of mass action, are inadequate for understanding or even for formulating the complexity induced by the evolution of nonlinear systems. In contrast attractors, fractals and multifractals, normal forms, Lyapunov exponents, entropies, invariant measures and correlation functions are parts of the new scientific vocabulary proposed by modern nonlinear science and provide a pragmatic way to meet a challenge in front of which classical approaches fail.

The aim of this book is to develop the material of an introductory course to nonlinear science which, while doing justice to its multiple facets, is not merely a compilation of topics but rather one (undoubtedly out of many) particular way to treat the entire field in a straightforward coherent manner. The book is addressed primarily to graduate level students and to researchers whose background is in physics, in applied mathematics, in chemistry or in engineering. In preparing it I tried to fill a gap that I have perceived in the literature ever since I began in the late 1970s–early 1980s to teach this subject to the fourth year students of physics and chemistry at the University of Brussels: while there are excellent books on particular aspects of nonlinear dynamics like bifurcation theory, chaos, fractals or ergodic theory, there is (with some rare exceptions) a marked tendency for uneven coverage of these subjects. Sometimes this takes extreme forms where nonlinearity is identified with chaos or on the contrary with everything but chaos; where difference equations and low-order systems dominate entirely over spatially extended systems; or where probabilistic aspects are either forgotten altogether or merely used as auxiliary material in the definition of average quantities such as attractor dimensions or Lyapunov exponents.

The book is addressed primarily to the practitioner of nonlinear science rather than to the theorem prover. The exposition is largely self-contained, but inevitably every now and then advanced knowledge from other fields is required. In such instances I have tried to appeal to common sense and to anticipate links with later chapters of the book. Moreover, despite repeated reference to and examples of conservative systems the book eventually carries its author's personal bias and interest in dissipative systems.

The principal role of the first two chapters is to motivate the development of the more technically oriented last five chapters. Chapter 1 provides an overview of the experimental evidence of nonlinear behavior in the physical sciences and biology thanks to which unifying concepts such as instability, bifurcation or symmetry-breaking are sorted out in view of later developments. In Chapter 2 the evolution equations corresponding to the systems and phenomena surveyed in Chapter 1 are laid down. A number of 'canonical' models used extensively later for illustrating the techniques of nonlinear science are also derived from these general equations. In Chapters 3 – 5 the techniques of nonlinear science are introduced on dynamical systems with a finite number of variables. We consider, successively, the geometry of phase space, the concepts of invariant manifold and of attractor, stability, and the bifurcation of new branches of solutions with special emphasis on normal form theory. Chapter 6 is devoted to spatially extended systems. It focuses on new problems such as pattern formation arising from the spatial degrees of freedom, and on the possible extensions of normal form theory. Chapter 7 deals with chaotic dynamics. This immense subject, to which tens (if not hundreds) of books have been devoted cannot, of course, be treated exhaustively in a single chapter. The particular approach I have chosen is to provide some classical introductory material and to focus for much of the remainder of the chapter on the statistical aspects of chaos, which usually are not covered adequately in the literature. I am convinced that these aspects along with spatio-temporal chaos introduced in the last section of the chapter will dominate research on chaos in the next years.

In preparing this book I have greatly benefitted from discussions with and the critical comments of my students, coworkers and colleagues. To the students of my nonlinear class I owe much of the choice of material and the 'experimental' proof that a 30-hour introductory course on nonlinear science of the kind I had in mind is possible. I am most indebted to C. Baesens with whom I had prepared an early (1981–2) version of my lecture notes entitled '*Phénomènes non-linéaires*'. C. Nicolis provided invaluable help in the preparation of the figures and in the final corrections. To the critical reading of I. Antoniou, P. Borckmans, D. Daems, T. Erneux and J. Weimar I owe the elimination of several misprints and mistakes as well as certain improvements of the first typed version that I circulated in December, 1992. Finally, it is a pleasure to thank S. Wellens and I. Saverino for the competent typing of successive versions of the manuscript and their patience, and P. Kinet for efficient technical assistance.

My research in the subject area covered in this book is sponsored by the University of Brussels, the Belgian Government, the Belgian Fund for Scientific Research and the European Commission. Their interest and generous support are gratefully acknowledged.

G. Nicolis
Brussels, December, 1993

CHAPTER ONE

Nonlinear behavior in the physical sciences and biology: some typical examples

1.1 What is nonlinearity?

Introductory science textbooks – and much of our educational system, for that matter – are built on the idea that a natural system subjected to well-defined external conditions will follow a unique course and that a slight change in these conditions will likewise induce a slight change in the system's response. Owing undoubtedly to its cultural attractiveness, this idea, along with its corollaries of reproducibility and unlimited predictability and hence of ultimate simplicity, has long dominated our thinking and has gradually led to the image of a *linear* world: a world in which the observed effects are linked to the underlying causes by a set of laws reducing for all practical purposes to a simple proportionality.

Appealing and reassuring as it may sound, this perennial idea is now being challenged and shown to provide, at best, only a partial view of the natural world. In many instances – and as a matter of fact in most of those interfering with our everyday experience – we witness radical, qualitative deviations from the regime of proportionality. This book has to do with *nonlinearity*, that is to say, the phenomena that can take place under these conditions.

A striking difference between linear and nonlinear laws is whether the property of superposition holds or breaks down. In a linear system the ultimate effect of the combined action of two different causes is merely the superposition of the effects of each cause taken individually. But in a nonlinear system adding two elementary actions to one another can induce dramatic new effects reflecting the onset of cooperativity between the constituent elements. This can give rise to unexpected structures and events whose properties can be quite different from those of the underlying elementary laws, in the form of abrupt transitions, a multiplicity of states, pattern formation, or an irregular markedly unpredictable evolution in space and time referred to as deterministic chaos. Nonlinear science is, therefore, the science of evolution and complexity.

By focusing on a specific class of behaviors encountered in many different contexts nonlinear science cuts across traditional scientific disciplinary divisions. It constitutes, today, one of the most active and rapidly growing branches of science.

The aim of this first chapter is to show how nonlinearities arise in a very broad range of natural phenomena, from classical mechanics to biology. At this early stage of our program the exposition will be qualitative and will appeal to undergraduate-level background knowledge and to common sense. In the subsequent chapters the tools of nonlinear science will be developed. This will gradually lead to a deeper understanding of the various phenomena touched upon in the present chapter.

1.2 Nonlinear behavior in classical mechanics

Nonlinear behavior is deeply rooted in the fundamental laws of classical physics. In this section we illustrate the ubiquity of nonlinearity in mechanics (Andronov, Vitt and Khaikin, 1966; Thompson, 1982), using the very simple example of the *hoop*. We consider (Fig. 1.1) a rigid vertical ring of radius r in the field of gravity. A mass m is initially placed at an angle θ_0 from the lower end of the vertical diameter and is allowed to move

Fig. 1.1 Schematic representation of the motion of a mass m on a vertical rotating hoop.

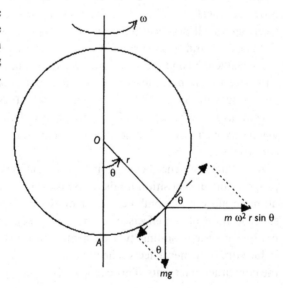

along the ring with no friction. As long as the ring as a whole is at rest, it will perform a periodic motion around position A (if $\theta_0 \neq 0$) or will remain fixed for ever on A (if $\theta_0 = 0$), the *equilibrium* state of our simple device.

Now let the ring be rotated around its vertical diameter with a constant angular velocity ω, as a result of an external *constraint* (here an appropriately applied torque). Experiment shows that as long as ω is small the mass m still oscillates around the same equilibrium position A as before. But, beyond a *critical threshold* ω_c, one observes that the situation changes completely and the mass oscillates around a new equilibrium position corresponding to a nonzero value of the angle θ. Actually there exist two such equilibria, placed symmetrically around the vertical diameter. There is no preference for either of the equilibria to be chosen: the choice is dictated by the initial position and velocity of the mass which in many respects is governed by chance. Still, in a given experiment only one of these equilibria will be realized and the mass will accordingly oscillate around it. To the observer, this will appear as an asymmetric realization of a perfectly symmetric physical situation. We refer to this phenomenon as *symmetry breaking*, the particular symmetry broken being here the reflection symmetry around the vertical diameter.

It is convenient to organize this information on a diagram (cf. Fig. 1.2) in which the equilibrium position θ, characterizing the state of our system, is plotted against the angular velocity ω – the constraint acting on the system. Below the threshold ω_c only one position is available, corresponding to $\theta = 0$ (branch (a) in Fig. 1.2). Beyond ω_c this state cannot be sustained. We express this in Fig. 1.2 by the dashed line along the branch (a') extrapolating branch (a). For each $\omega > \omega_c$ two new equilibria become available. Joining the corresponding values of the angles we obtain two branches of states (b1) and (b2) which merge with (a) at $\omega = \omega_c$ but

Fig. 1.2 Bifurcation of new equilibria θ_+, θ_- as the angular velocity ω of the hoop exceeds the threshold value ω_c.

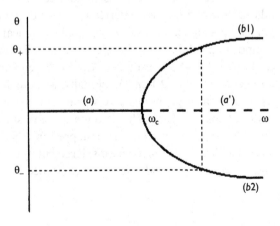

separate from it at $\omega \neq \omega_c$. This is the phenomenon of *bifurcation*, which will recur throughout this volume and which will turn out to be one of the most characteristic signatures of nonlinearity.

In summary, we have seen using a very simple mechanical device that beyond a critical threshold of constraint nonlinear phenomena are switched on in the sense that the system responds to the constraint in a manner that deviates dramatically from the law of proportionality referred to in Section 1.1. One of the manifestations of this nonlinear response is that the system now disposes of multiple solutions between which it can choose. As the value of ω gradually grows we switch spontaneously from the classical regime of pendulum-like oscillations around a unique equilibrium to a regime of oscillations around a variety of possible equilibria. It is significant that the same – quite ordinary – physical system can present different types of behavior as the values of a characteristic parameter built into it are varied.

It is easy to figure out the qualitative mechanism at the root of this phenomenon. The motion of the mass m is governed by two adverse factors (Fig. 1.1): the weight mg that tends to move it downwards, and the centrifugal force $m\omega^2 r \sin \theta$ that tends to maintain it away from the downward vertical position A. Intuitively, the higher the ω the stronger the tendency to sustain a motion away from A will be. This will be substantiated by the more quantitative formulation of the next chapter.

As we shall see shortly, the above features are not limited to our simple example but are prototypical of a large class of natural systems. In the case of the hoop it so happens that they constitute the exhaustive list of the types of complex behavior that become possible in the nonlinear range. But this is by no means a universal limitation. The action of an external periodic forcing on a nonlinear oscillator of the kind depicted in Fig. 1.1 (through, for instance, a periodic variation of the radius, $r = r_0 + r_1 \sin \Omega t$) already gives rise, under certain conditions, to an aperiodic motion of the mass m referred to in Section 1.1 as *deterministic chaos*. Similar phenomena arise in coupled nonlinear oscillators like the composite pendulum and in celestial mechanics in connection with the celebrated three-body problem. A more systematic classification will be outlined in Sections 2.1 and 3.3. Finally, we note that electric circuits give rise to nonlinear phenomena very similar to those arising in mechanical devices (Linsay, 1981). This is quite natural, since the laws governing these two types of system can be mapped into each other provided that mass is replaced by inductance, displacement by charge, restoring force by inverse capacitance and friction by resistance.

1.3 Thermal convection

We now turn to macroscopic physics and survey nonlinear behavior in connection with *thermal convection* – the bulk motions of fluids generated by temperature inhomogeneities.

Thermal convection is at the origin of important, spectacular natural phenomena. Some examples are the circulations of atmosphere and oceans, which determine to a large extent short- and medium-term weather changes and continental drift, the motion of continental plates induced by large-scale movements in the mantle. In this section we shall be interested in the more modest laboratory-scale experiment first performed in 1900 by Bénard which, despite its apparent simplicity, leads to the observation of a number of astonishing properties (Chandrasekhar, 1961; Koschmieder, 1981; Velarde and Normant, 1980).

Imagine a thin fluid layer between two horizontal conducting plates in the field of gravity (Fig. 1.3). The plates are maintained at fixed temperatures T_0 and T_1, T_0 being larger than or equal to T_1, through appropriate heating from below. Suppose first that $\Delta T = T_0 - T_1 = 0$. The layer will then sooner or later reach the state of thermodynamic equilibrium, characterized by the absence of bulk motion and by a uniform temperature and density throughout. Now let a temperature difference $\Delta T = T_0 - T_1 > 0$ be gradually applied. This *thermal constraint*, the analog of the mechanical constraint generating the angular velocity ω in the example of the previous section, displaces the system from equilibrium and gives rise to heat conduction from the lower (hot) plate to the upper (cold) one and a concomitant temperature distribution along the vertical. As long as ΔT is weak the behavior will be limited to this. In particular, the fluid will remain at rest and an observer moving along a horizontal plane will still perceive a uniform temperature and density environment.

The situation changes completely when ΔT exceeds a *critical threshold* ΔT_c. The fluid ceases now to be at rest and begins to perform bulk movement organized in the form of well-structured convection cells (Fig.

Fig. 1.3 Schematic representation of a horizontal fluid layer heated from below.

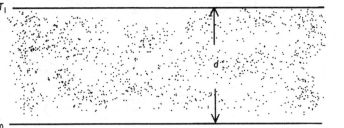

1.4), known as Bénard cells (this is precisely the regime of thermal convection). In a given cell the fluid moves upward, follows the upper boundary, then sinks downward, follows the lower boundary, and starts all over again. If this motion occurs, say, in the clockwise direction in the adjacent cells it will occur in the opposite, counterclockwise direction. The cells have a characteristic size which is determined once the geometry – and particularly the depth d – is specified. Notice that despite the fact that the fluid moves, the velocity, temperature and density at a given point are time-independent. From the macroscopic point of view, therefore, we have a stationary nonequilibrium state.

We have seen that two adjacent cells rotate in opposite directions. At a given point in space, therefore, a small volume element of the fluid can find itself at two distinct states in the sense that it can be part of a cell rotating clockwise or counterclockwise. Actually, since for a system of large horizontal extent the whole structure can be shifted by any amount along the horizontal direction there is a whole continuum of states available. But since the most characteristic manifestation of this multiplicity remains the direction of rotation of the cell, we will argue in the following in terms of a two-fold multiplicity. Now, nothing in the experimental setup allows one to assign beforehand a preference for either of these two directions: the particular direction that will be chosen at a particular point of our apparatus will merely be dictated by the locally prevailing conditions at the moment of the experiment which, to a large extent, are determined by random elements such as minute local temperature fluctuations, dust particles and mechanical vibrations. Still, in a given experiment, a volume element of the fluid located around this particular point will eventually realize only one of these two types of motion. Just like in Section 1.2, we witness here an asymmetric realization of an initially perfectly symmetric physical situation. This symmetry-breaking phenomenon is associated with the breaking of the chiral symmetry

Fig. 1.4 Qualitative view of convection (Bénard) cells.

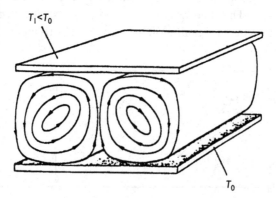

$T_1 < T_0$

T_0

associated with the sense of rotation. Alternatively, for an observer moving along the horizontal plane homogeneity will be broken: since there is internal spatial differentiation within a pair of adjacent cells, to recover the same environment one must now move a distance twice the size of the cell. In other words the Bénard convection also breaks the translational symmetry in the horizontal direction. The situation is somewhat analogous to a liquid–solid transition, where the full isotropy and rotational symmetry of the liquid phase are broken in favor of the less symmetric crystalline solid. A major difference is that in a crystal the characteristic length associated with translational symmetry breaking is microscopic, comparable to the range of intermolecular forces. In contrast the characteristic size of a Bénard cell is macroscopic, of the order of a millimeter or more. In some sense one may think of such large-scale patterns as macroscopic, 'dissipative' crystals of a completely new kind. Just like their equilibrium analogs, such structures can also be classified according to their spatial symmetries (Walgraef, Dewel and Borckmans, 1982; Manneville, 1991). One may thus expect structures belonging to the cubic, rhombohedral or tetragonal systems, many of which are indeed observed in thermal convection experiments.

Fig. 1.5 shows a qualitative explanation of the phenomenon using, just like in Section 1.2, the competition between two adverse factors. Owing to thermal expansion the fluid becomes stratified, with the part close to the lower plate characterized by a lower density than the upper part. This gives rise to a density gradient that opposes the force of gravity – a potentially unstable configuration. Consider a small volume of the fluid near the lower plate. Imagine that this volume element is slightly displaced upward by the random action of disturbances that are inevitably acting on any real-world system. Being now in a colder – and hence denser – region it will experience an upward force that will tend to amplify further the ascending movement. If, on the other hand, a small droplet initially close to the upper plate is displaced downward, it will penetrate an environment of lower density and the initial descent will be further amplified. We see therefore that, in principle, the fluid can generate

Fig. 1.5 Qualitative explanation of the origin of thermal convection.

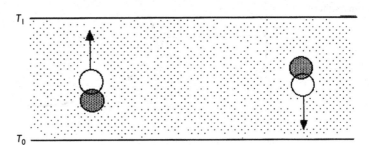

ascending and descending currents like those observed in the experiment. The reason these currents do not appear as soon as ΔT is not strictly zero is that the destabilizing effects are counteracted by the stabilizing effects of viscosity, which generates an internal friction opposing movement, as well as by thermal conduction, which tends to smear the temperature difference between the displaced drop and its environment. This explains the existence of the critical threshold ΔT_c observed in the experiment.

In analogy, once again, with Section 1.2 we may organize the above information in the form of a *bifurcation diagram* (Fig. 1.6). A convenient state variable to be plotted against the constraint ΔT is now the vertical component of the velocity, w, at a given point, say in the middle of the layer. Before the threshold ΔT_c the fluid is at rest, $w = 0$. Beyond ΔT_c at the particular point under consideration one will observe an ascending ($w > 0$) *or* a descending ($w < 0$) movement. These two branches of states coalesce at ΔT_c with the state of rest, but bifurcate out of it for $\Delta T > \Delta T_c$. Despite the completely different natures of the systems considered in Sections 1.2 and 1.3 we sense here some unity: in both cases nonlinear behavior, associated with multiplicity of states, emerges through a bifurcation mechanism when a constraint acting on the system exceeds a critical threshold, the states born from bifurcation being qualitatively different from the state prevailing before bifurcation by the fact that they display broken symmetries. As will become gradually more and more evident in this book this scenario is typical and underlies huge classes of nonlinear phenomena which arise in widely different contexts.

One of the reasons that make the Bénard problem so important in nonlinear science is that, in addition to this first bifurcation, the system can also undergo a whole series of successive transitions unveiling practically the entire repertoire of nonlinear behaviors known to date. Several transition scenarios have been discovered, thanks to the use of

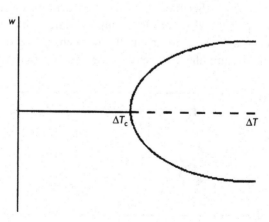

Fig. 1.6 Bifurcation diagram for the onset of thermal convection beyond a critical temperature difference ΔT_c.

increasingly sophisticated optical techniques. As it turns out, they depend on two main factors: the *aspect ratio* $\Gamma = L/d$, measuring essentially the horizontal extent L of the system; and the fluid's intrinsic parameters such as viscosity or heat conductivity. Typically, for Γ around 2 or so (a 'small' system), when ΔT is increased to a value several times larger than ΔT_c, say ΔT_{c_2}, convection is no longer steady but becomes periodic in time. The frequency of the oscillation is in the macroscopic range and is intrinsically determined by the system (Dubois and Bergé, 1981). Beyond a still higher threshold ΔT_{c_3} a second frequency appears and the behavior becomes biperiodic. Two cases may now be distinguished. In the first one the two frequencies ω_1 and ω_2 have a common divisor, say Ω. In practical terms this is guaranteed once their ratio ω_1/ω_2 is rational, that is to say, equal to the ratio of two integers p and q: $\omega_1/\omega_2 = p/q$. The common divisor Ω is then $\Omega = \omega_1/p = \omega_2/q$ and plays the role of the frequency of a global, simply-periodic motion. In the second case this condition is not fulfilled, a fact that one usually summarizes by the statement that ω_1 and ω_2 are 'irrationally related'. The doubly-periodic regime is, then, *quasi-periodic*. Since the irrationals constitute the overwhelming majority of real numbers, one expects that this latter case will be the most typical one to be realized in the experiment. Finally for a temperature difference ΔT beyond a new threshold ΔT_t the time dependence becomes *chaotic*. In the time domain this shows up through aperiodicity of a more noisy type than quasi-periodic behavior, despite the deterministic character of the underlying system. This type of behavior is reminiscent of turbulence.

A very convenient quantitative distinction between periodic, quasi-periodic and chaotic regimes is provided by the *power spectra* constructed by Fourier transforming the time series data of a relevant variable $X(t_k)$, $k = 1, \ldots, N$ (Bergé, Pomeau and Vidal, 1984):

$$P(\omega_j) = |\tilde{X}(\omega_j)|^2 = \left| \frac{1}{N^{1/2}} \sum_{k=1}^{N} e^{-i\left(\frac{2\pi k}{N}\right)j} X(t_k) \right|^2 \tag{1.1}$$

where $\omega_j = j/(N\Delta t), j = 0, \ldots, N-1$ and $\Delta t = t_k - t_{k-1}$. For periodic or quasi-periodic behavior the power spectrum consists of only isolated lines of well-defined frequencies, whereas chaotic behavior is marked by a broad band continuous spectrum containing an important low-frequency part. Such a transition from periodicity to chaos in the Bénard experiment is depicted in Fig. 1.7.

It should be noted that a broad band spectrum can be compatible with both (deterministic) chaos and (random) noise. There exist more elaborate ways than power spectra, allowing one to characterize chaos and to discriminate it from both periodicity and random noise. They will be developed later on in this book, when the geometric and analytic techniques of nonlinear science will be laid down (see especially Chapters 3 and 7).

The quasi-periodic route to chaos established by the above experiments has a historical significance since it provides a refutation of the ideas prevailing until the 1970s on the nature of turbulence. Specifically, it was thought that turbulence arises from an infinity of transitions, each generating a new frequency until a continuum of such frequencies characteristic of turbulent spectra becomes available. The possibility that turbulent-like behaviors may arise after a finite (and small) number of

Fig. 1.7 Fourier spectra drawn from velocity measurements in the Bergé–Dubois experiment of the Rayleigh–Bénard instability (Dubois and Bergé, 1981): (*a*) periodic, (*b*) quasi-periodic and (*c*) weakly chaotic regimes. In (*b*), measurement at two different points reveal two different, irrationally related frequencies. R_c is a dimensionless measure of the critical temperature difference ΔT_c.

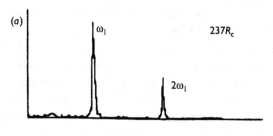

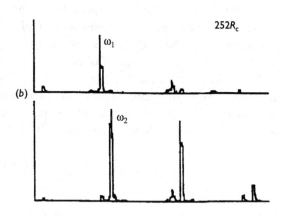

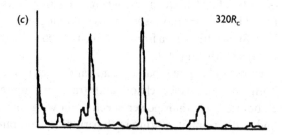

transitions, first proposed theoretically by Ruelle and Takens (1971) before the experimental confirmation outlined above, sheds therefore new light on this major problem. However, one should be aware of the fact that the chaotic behavior observed under these conditions is far from the fully developed turbulence characterizing large-scale real-world flows.

Two other transition to chaos scenarios can be observed in small systems under different conditions. In the so-called intermittency route, after a first transition to a time-periodic flow one observes a further transition to a chaotic state characterized by long quiescent periods of nearly periodic behavior interrupted, at more or less random intervals, by short-lived bursts (Bergé *et al.*, 1984). In the so-called period doubling cascade, after a first transition to simple periodic behavior a long series of transitions to complex periodic oscillations is observed in which the main frequency is a subharmonic of the previous one. Eventually chaotic behavior sets in (Libchaber and Maurer, 1980). These scenarios will be discussed more fully in Chapter 7.

Let us turn now to systems of large horizontal extent, characterized by an aspect ratio which is much larger than unity (Ahlers and Behringer, 1978; Pocheau, Croquette and Le Gal, 1985). A transition to a chaotic convection is again observed, but there are some major differences with the behavior observed in small systems. First, the successive transition thresholds are now squeezed into a small vicinity near the threshold of stationary thermal convection, ΔT_c. Second, in addition to the weakly chaotic behavior typical of small systems 'hard' turbulence becomes possible. Third, and perhaps most significant of all, the convective motion is ordered only on a local scale. Specifically, it appears (Fig. 1.8) that

Fig. 1.8 Convection patterns in a system of large aspect ratio (courtesy of Croquette and Pocheau). When roll-like structures of different orientations merge defects of various kinds (dislocations, grain boundaries . . .), indicated by the arrows, are formed.

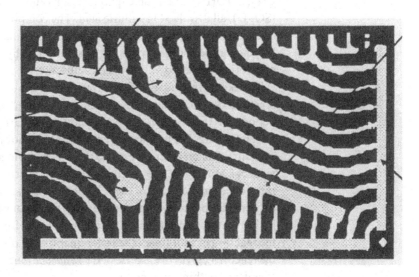

different patterns (corresponding to different bifurcation branches in Fig. 1.6) are realized in different parts of the system. When they merge *defects* are created and subsequently undergo a complex dynamics which appears to be the prelude to spatio-temporal chaos, a regime in which in addition to aperiodicity and irreproducibility in the time domain one also observes an erratic distribution of hydrodynamic variables in space.

The behavior found in the Bénard experiment turns out to be typical of a large class of phenomena arising in fluid dynamics (Chandrasekhar, 1961; Lin, 1955; Guyon, Hulin and Petit, 1991) such as the motion of a fluid between two concentric rotating cylinders (Taylor vortex flow), thermal convection in a mixture (Soret flow), and transitions in the presence of a free surface (Bénard–Marangoni convection). In the following, the Bénard problem will be one of the typical models to which we will refer to illustrate the ideas and tools that we shall put forward to understand nonlinear phenomena.

1.4 Nonlinear phenomena in chemistry

For a long time chemists thought that a homogeneous, time-independent state should eventually emerge from any chemical transformation. The first clearcut experimental evidence that this view is incorrect came in the 1960s from a redox reaction known as the Belousov–Zhabotinski (BZ) reaction (Zhabotinski 1964; Nicolis and Portnow, 1973; Noyes and Field, 1974).

A typical BZ reagent consists of such ordinary products as a salt which generates bromate (like $KBrO_3$), an organic reductant (like malonic acid $CH_2(COOH)_2$) and a salt capable of generating a redox couple (like $Ce_2(SO_4)_3$), all dissolved in sulfuric acid. The composition of the system can be followed visually through a change in color and, more quantitatively, by placing specific electrodes in the solution or by measuring the optical absorption caused by a particular substance.

Thanks to the design of open reactors (Pacault and Vidal, 1978) the reaction can nowadays be carried out without interruption for long periods of time – a necessary condition to arrive at a quantitative understanding. We first survey briefly the behavior of the BZ reaction in a well-stirred open reactor (Fig. 1.9) in which homogeneity is ensured by a vigorous stirring of the mixture, hereafter referred to as 'CSTR'. Two types of parameter control the behavior of this system: the concentrations of chemicals pumped from outside, and the rate at which they are pumped in the reactor, that is to say, the volume pumped from the feed stream into the reactor per unit time. The latter quantity J_i, divided by the volume of the reactor V, gives the inverse of the *residence time* τ_i of the corresponding chemical i within the reactor,

$$1/\tau_i = J_i/V \qquad (1.2)$$

Under slow pumping conditions (all J_i small) the chemicals will remain in the reactor for a very long time and will, for all practical purposes, reach a state of chemical equilibrium just as if the reactor were closed. However, for large J_i the chemicals will leave the reactor very quickly, essentially with the concentration of the feed stream and will be unable to react significantly and equilibrate with the bulk. The residence time τ can therefore be used as a convenient control parameter playing the role of the constraint, much like the angular velocity ω in Section 1.2 or the temperature difference ΔT in Section 1.3.

We now survey the principal modes of behavior of BZ-like reactions in an open well-stirred reactor for a range of values of the residence time between the two above-mentioned extremes (Vidal and Pacault, 1981; Bergé *et al.*, 1984). As it turns out, there exists a critical threshold τ_c below which stationary behavior is no longer possible and sustained oscillations are observed. The amplitude and period of these oscillations are intrinsically determined by the dynamics once the parameters (temperature, concentrations in the feed streams and residence times) are specified. The birth of this new regime is associated with the breaking of translational symmetry in the time domain, since then the phase of the system's variables changes within an oscillation period. As in Sections 1.2 and 1.3, the transition can be represented in the form of a bifurcation

Fig. 1.9 Experimental setup for an open chemical system.

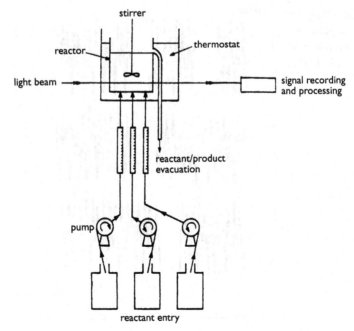

diagram in which the oscillatory branch bifurcates out of the stationary state when $\tau = \tau_c$.

Fig. 1.10 depicts a sequence of transitions leading to complex periodic oscillations of various types, as the residence time is varied (Simoyi, Wolf and Swinney, 1982). Further change of the constraints leads to a regime of chemical chaos, in the form of a random-looking mix of small- and large-amplitude oscillations. As in the thermal convection problem, the chaotic regime can be reached in a variety of different ways depending on the values of the constraints, including the period doubling cascade, a transition through quasi-periodic behavior and intermittency.

A qualitative explanation of the oscillatory behavior in the BZ reagent was proposed by Field, Körös and Noyes, (1972). Although their original mechanism has been modified over the years, it still provides the basis of our knowledge of this system. The main point to realize is that the BZ reaction involves two different processes, A and B, which alternately dominate the kinetics, while a third process C is responsible for a switching from B to A. More specifically:

- When the Br^- concentration is appreciable, the following reactions take place,

Fig. 1.10 Complex periodic oscillations in BZ reaction (Simoyi *et al.*, 1982).

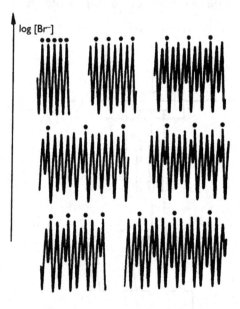

$$(A.1) Br^- + BrO_3^- + 2H^+ \rightarrow HOBr + HBrO_2$$
$$(A.2) Br^- + HOBr + H^+ \rightarrow Br_2 + H_2O$$
$$(A.3) Br^- + HBrO_2 + H^+ \rightarrow 2HOBr$$

- Once Br^- concentration is sufficiently lowered, the oxidation of Ce^{3+} to Ce^{4+} is carried out autocatalytically thanks to the liberation of the free radical $BrO_2\cdot$,

$$(B.1) 2HBrO_2 \rightarrow HOBr + BrO_3^- + H^+$$
$$(B.2) HBrO_2 + BrO_3^- + H^+ \rightarrow 2BrO_2\cdot + H_2O$$
$$(B.3) BrO_2\cdot + Ce^{3+} + H^+ \rightarrow Ce^{4+} + HBrO_2$$

Autocatalysis takes place through the bromous acid, since two molecules of this substance are produced by $(B.3)$ for each one consumed in $(B.2)$.

- Finally, Ce^{4+} is reduced back to Ce^{3+} while at the same time it regenerates Br^-, thus allowing the process to start all over again. The global reaction for this transformation can be written as

$$(C) 10Ce^{4+} + CH_2(COOH)_2 + BrCH(COOH)_2 + 4H_2O + 2Br_2$$
$$\rightarrow 10Ce^{3+} + 5Br^- + 6CO_2 + 15H^+$$

Notice that the mere fact of switching from pathway B to pathway A through process C is not sufficient to explain oscillations. An additional ingredient – cooperativity – is needed to sustain a cyclic behavior, and is provided by the positive feedback (autocatalysis) of bromous acid onto itself. We may express this schematically through the following diagram:

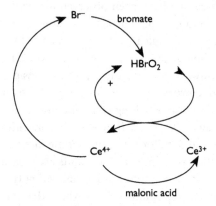

Being limited to a single feedback, the above mechanism cannot generate more than one relevant time scale. Consequently, it cannot explain the quasi-periodicity, composite periodic oscillations and chaos found in numerous experiments. Richetti and coauthors have proposed an augmented Field – Körös – Noyes mechanism involving an additional, negative feedback whereby in the presence of hypobromous acid (HOBr), $HOBr_2$ is consumed in a manner that is autocatalytic in Br^-. The competition between these two mechanisms of cooperativity then brings a second time scale into the problem. This suffices to explain, in principle, the appearance of complex oscillations including chaotic ones (Argoul, Arnéodo, Richetti and Roux, 1987; Arnéodo, Argoul, Elezgaray and Richetti, 1993).

We now come to the behavior of the BZ system in an unstirred reactor (Vidal and Pacault, 1984; Field and Burger, 1985; Gray, Nicolis, Baras, Borckmans and Scott, 1990). For a long time experiments were limited to closed reactors, typically in the form of a shallow layer of solution contained in a petri dish. Fig. 1.11 depicts the typical behavior found under these conditions, namely regular patterns in space and time in the form of propagating wave fronts. The waves appear primarily in two different forms: circular fronts (*a*) displaying a roughly cylindrical symmetry around an axis perpendicular to the layer, usually referred to as *target patterns*; and spiral fronts (*b*) rotating in space clockwise or counterclockwise. It is also possible to obtain, although under rather exceptional conditions, the multiarmed spirals shown in Fig. 1.11(*c*). In each case the wave fronts propagate over macroscopic distances without distortion and at a prescribed speed. As in the Bénard problem we can associate the formation of these fronts with space symmetry breaking: translational symmetry breaking in the target wave case and chiral symmetry breaking in the spiral wave case.

More recently *stationary* inhomogeneous patterns arising through symmetry breaking have been observed in a variant of the BZ reaction, the chlorite – iodide – malonic acid reaction, thanks to the design of new open, unstirred chemical reactors which allow the development of spatially inhomogeneous states while avoiding parasitic hydrodynamic motion. Figure 1.12 depicts the gel reactor designed to that effect by De Kepper and coworkers (Castets, Dulos, Boissonade and De Kepper, 1990). The two opposite long edges are in contact respectively with two chemical reservoirs, *A* and *B*, where the concentrations of reactants are kept constant and uniform by appropriate mixing and a continuous flow of fresh reactant solutions. The reactants are separated in such a way that neither solution *A* nor solution *B* is individually reactive: the chemicals diffuse from the edges into the gel where the reaction takes place. The typical diffusion time to establish stationary concentration profiles across the gel strip is of the order of an hour. To make the concentration changes visible, the gel is loaded with a starch-like color indicator which does not diffuse through the gel. The color changes from yellow to blue with the change of the $[IO_3^-]/[I_2]$ ratio during the redox reaction. The color pattern is monitored with a video camera.

Fig. 1.11 Wave
propagation in a
two-dimensional layer
of BZ reagent: (*a*)
target patterns; (*b*)
spiral waves; (*c*)
multiarmed spirals.

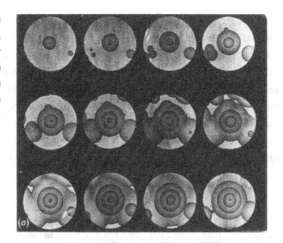

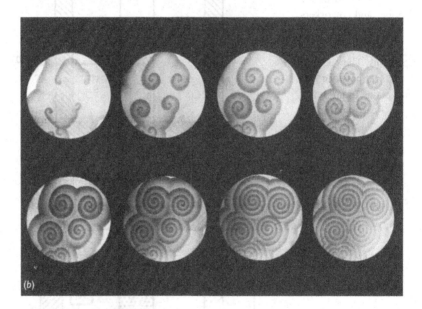

At the beginning of the experiment, the development of a series of light and dark stripes parallel to the edges reveals the emergence of a concentration pattern in the central region of the reactor. Although this pattern is nontrivial, the stripes preserve the symmetry imposed by the feed. But over a well-defined range of the malonic acid concentration in *B*, some of these stripes ultimately break up into lines of periodic spots as depicted in Fig. 1.13. This constitutes a genuine symmetry-breaking phenomenon in the direction transverse to the imposed gradient. The pattern can be sustained indefinitely. Moreover, the wavelength ($l_c \approx 0.2$ mm) seems to be intrinsic and exclusively characterized by

Fig. 1.12 The open unstirred reactor designed by De Kepper and coworkers for the realization of stationary spatial patterns. The gel strip is fixed between two flat plates 1 mm apart. Reactants are fed through the well-mixed reservoirs *A* and *B*.

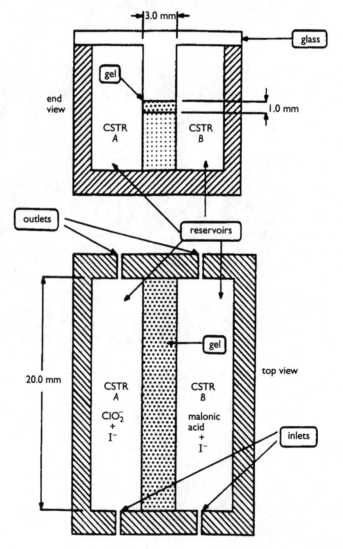

nongeometric properties; in particular, it is much smaller than any geometric size of the reactor (including thickness) by at least one order of magnitude. This is very different from the stationary patterns in the Bénard problem, whose characteristic length is determined by the size of the experimental device.

In the above experiment, visualization of the pattern is perpendicular to the direction of feed. An alternate possibility is visualization parallel to the direction of feed, which is, in turn, chosen to be perpendicular to the reactor's largest dimension (Ouyang and Swinney, 1991). Fig. 1.14 reveals in this case a very large variety of patterns such as hexagons, stripes and mixed states. Transitions from stationary spatial patterns to spatio-temporal chaos (chemical turbulence) are also observed in conjunction with the formation of defects separating different domains, much as in the Bénard problem. These phenomena will be analyzed in detail in Chapters 6 and 7.

1.5 Some further examples of chemically-mediated nonlinear behavior.

In addition to their fundamental interest in laboratory-scale chemistry, chemically-mediated nonlinearities provide the natural explanation of a large body of complex phenomena in a wide variety of contexts. In this section we survey some characteristic examples from the fields of chemical engineering and biology (see e.g. Nicolis and Baras (1984), Gray *et al.* (1990) and Murray (1989)).

Much of the chemical industry is based on heterogeneous catalysis. In this, some of the steps necessary for the synthesis of a product are accelerated by the presence of a surface on which the chemicals are first adsorbed and then converted to active forms capable of undergoing reactions that would be impossible otherwise. For instance, the oxidation of ammonia or of carbon monoxide is usually carried out in the presence of a platinum catalyst; similarly in the decomposition of nitrous oxide, N_2O, a catalytic copper oxide surface is used.

There are two quite general sources of cooperativity in this type of

Fig. 1.13 Enlarged image of the region of the reactor, Fig. 1.12, in which the pattern is appearing (Castets *et al.*, 1990). Distances are in millimeters. Dark regions correspond to reduced states, light ones to oxidized states.

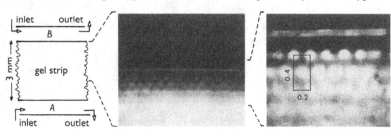

phenomenon. First, since the total number of sites in the catalyst is finite, the substances in the bulk phase inevitably compete for adsorbing sites: they thus exert a negative feedback on each other. Second, some of the adsorbed substances facilitate a structural change of the catalyst surface, which may further affect the adsorption probability of this very substance or of other substances. It is, therefore, not surprising that heterogeneous catalysis provides today some of the best-documented examples of nonlinear phenomena, from homogeneous sustained oscillations to propagating waves or even to stationary concentration patterns. A particularly clear illustration is provided by the beautiful experiments of carbon monoxide oxidation on platinum under ultra-high vacuum ($< 10^{-3}$ mbar) pressure conditions (Cox, Ertl and Imbihl, 1985; Ertl, 1991; Imbihl, 1992).

Combustion, the burning of hydrocarbons, the process by which heat engines function, is another class of important chemical transformations. The overall reaction can be represented schematically as

Fig. 1.14 Stationary chemical patterns formed in a continuously fed unstirred reactor (Ouyang and Swinney, 1991). Visualization is parallel to the direction of feed: (*a*), (*b*) hexagons; (*c*) stripes; (*d*) mixed state.

$$\text{fuel} + \text{oxygen} \xrightarrow{k} \text{oxide} + \text{heat} \qquad (1.3)$$

Now, according to the well-known Arrhenius law of chemical kinetics, the rate constant k of a reaction is an increasing function of temperature (Kondratiev and Nikitin, 1981)

$$k(T) = k_0 \, e^{-E/RT} \qquad (1.4)$$

where E represents the activation barrier that the kinetic energy of the reactants must overcome in order to break the chemical bond. In an exothermic reaction like (1.3), the heat liberated will increase the translational kinetic energy and thus the temperature; this will increase the rate constant $k(T)$, eq. (1.4), and thus the rate of production of heat; this, in turn, will increase T and $k(T)$ further, whereupon more heat will be liberated, etc. This thermal feedback, which is ubiquitous in all combustion phenomena, is responsible for a rich variety of typically nonlinear behaviors, the most obvious example of which is the appearance of flame fronts separating the region of fresh reactants from the region of burnt ones.

One of the typical manifestations of nonlinear behavior in biology is self-sustained oscillations (Goldbeter, 1990; Segel, 1984; Murray, 1989). They are observed at all levels of biological organization, from the molecular to the supercellular, or even to the social (population) one, with periods ranging from seconds to days to years. Among these the best understood are biochemical oscillations at the subcellular level. Their most characteristic feature is that they involve enzyme regulation at a certain stage of the reaction sequence. The enzymes responsible for regulation are usually not simple Michaelian enzymes but, rather, *cooperative* (allosteric) ones in the sense that their conformation is affected by the fixation of certain metabolites and subsequently influences the catalytic activity of the enzyme. This cooperativity introduces, precisely, the nonlinearity necessary for complex behavior.

Another common manifestation of nonlinearity in biology is the coexistence of multiple steady states (see e.g. Thomas and d'Ari (1990)). Two very interesting contexts in which this behavior is likely to be manifested are the functions of the nervous and the immune systems, where it is thought to provide a prototype of the phenomenon of *memory* (Kaufman and Thomas, 1987). A less spectacular, but well-documented example is the ability of microorganisms to switch between different pathways of enzyme synthesis according to the medium in which they are embedded (Jacob and Monod, 1961). A source of nonlinearity common to all these phenomena is the almost stepwise response of the biomolecules to various effectors or even to their own or to the other unit's activity.

A most appealing example of symmetry breaking in living systems is morphogenesis. In the course of embryonic development one witnesses a sequence of events leading from a unique cell, the fertilized egg, to a multicellular organism involving specialized cells organized in an enormous variety of shapes and forms. Significantly, in developing tissues one frequently observes gradients of a variety of substances such as ions or relatively small metabolites. It has been conjectured that such gradients provide the tissue with a kind of 'coordinate system' that conveys *positional information* to the individual cells, by means of which they can recognize their position with respect to their partners and differentiate accordingly (Wolpert, 1969). Chemical symmetry-breaking bifurcations (cf. Fig. 1.13–1.14) provide an appealing prototype for understanding these processes. Further examples of space-dependent nonlinear behavior in biology include the propagation of a nerve impulse, the calcium-induced propagating waves on cell membranes and the peculiar aggregates formed by unicellular organisms like amoebae or bacteria.

For a long time it was thought that chaos in biology is tantamount to nuisance. A logical consequence of this attitude was to view it as the natural reference for understanding certain forms of biological disorder. This idea has been implemented in a number of convincing examples related, for instance, to respiratory diseases or to arrhythmias of the cardiac muscle (Glass and Mackey, 1988). Today it is realized that beyond this aspect chaos is likely to play a constructive role of the utmost importance in the highest levels of biological organization, particularly brain activity (J. S. Nicolis, 1991; Babloyantz and Destexhe, 1986; Destexhe, 1992). Consider as an example the electroencephalogram (EEG), a widely used record of the electrical activity of the brain generated by the sum of elemental sustained low-frequency (0.5–40 Hz) neuronal activities emanating from small volumes of cortical tissue just underneath the scalp (Fig. 1.15). The upper five panels of the Figure describe the EEG of a normal human subject in two stages of awareness ((*a*)–(*b*)) and three stages of sleep ((*c*)–(*e*)), whereas the lower two describe the EEG associated to two pathological situations. All records show an irregular succession of peaks, although (*f*) and (*g*) look definitely more 'coherent' than (*a*)–(*e*). The analysis of the time series associated with these records using the techniques of nonlinear dynamics developed later in this book reveals the presence of deterministic chaos, whose complexity depends on the stage of brain activity. Significantly, in pathological states such as epilepsy chaotic behavior is milder than in the healthy state. This suggests the rather unexpected idea that a healthy physiological system needs a certain amount of internal variability whose loss, witnessed by the

Fig. 1.15 EEG records . during various stages of brain activity (courtesy of A. Destexhe).

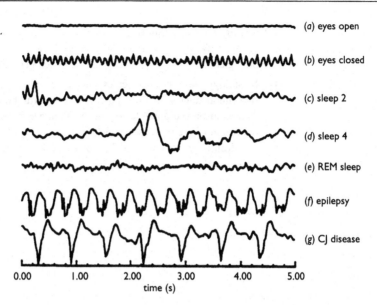

(a) eyes open

(b) eyes closed

(c) sleep 2

(d) sleep 4

(e) REM sleep

(f) epilepsy

(g) CJ disease

0.00   1.00   2.00   3.00   4.00   5.00

time (s)

transition to a more 'ordered' state, signals the appearance of pathological behavior. It is important to realize that this healthy variability is not the result of random noise. Deterministic chaos combines here, in a subtle manner, order and reliability on the one side, and disorder and unpredictability on the other. One is therefore tempted to speculate that a regime of deterministic chaos provides a system with the readiness needed to recognize a variety of external inputs and respond flexibly to a changing environment.

Population biology provides another example in which chaos may well be present. Experimental data are not clearcut, mainly because of the large environmental noise that is inevitably superimposed on the dynamical behavior. Still, taken in conjunction with mathematical models, they do suggest that irregular variations in time and space are ubiquitous (May, 1974, 1976). Such variations may have an important role in problems involving competition and selection.

Problems

1.1 Using eq. (1.1) show that the power spectrum of a sinusoidal signal displays a sharp peak around the signal frequency. Comment on the origin of the fine structure around this peak, particularly on its relation to the number of data points N. What happens for a nonsinusoidal periodic function in the form of a square pulse of width Δ repeated every T time units?

1.2 Identify the nonequilibrium constraint(s) analogous to ΔT of the Bénard problem (Section 1.3) and τ of the CSTR (Section 1.4) driving the biological rhythmic or patterning phenomena surveyed in Section 1.5.

1.3 In biology many key molecules, such as proteins, and many key subcellular microstructures possess a systematic built-in handedness. Comment on the possible role of this handedness at the molecular level in the origin of the macroscopic left–right asymmetry observed in vertebrates (Almirantis and Nicolis, 1987; Brown and Wolpert, 1990).

CHAPTER TWO

Quantitative formulation

2.1 Evolution equations in classical mechanics

The state variables of a mechanical system are the spatial coordinates $\{\mathbf{r}_i\}$ and velocities $\{\mathbf{v}_i\}$, or the generalized positions $\{\mathbf{q}_i\}$ and momenta $\{\mathbf{p}_i\}$, $i = 1, \ldots, N$, of the N constituent particles. These quantities vary in space and time owing to the interactions between the particles and/or to the forces of external origin acting on them. The evolution laws are Newton's equations,

$$m_i \frac{\mathrm{d}^2 \mathbf{r}_i}{\mathrm{d}t^2} = \mathbf{F}_i(\{\mathbf{r}_j\}) \quad i = 1, \ldots, N \tag{2.1}$$

or Hamilton's equations (Goldstein, 1959)

$$\left. \begin{aligned} \mathrm{d}\mathbf{q}_i/\mathrm{d}t &= \partial H/\partial \mathbf{p}_i \\ \mathrm{d}\mathbf{p}_i/\mathrm{d}t &= -\partial H/\partial \mathbf{q}_i \quad i = 1, \ldots, N \end{aligned} \right\} \tag{2.2}$$

in which m_i stands for the mass of ith particle and H for the Hamiltonian. Eqs. (2.1) and (2.2) share the generic property of *nonlinearity*, since the fundamental laws of nature depend on the particle coordinates in a nonlinear fashion, a typical example of which is Newton's law of gravitation.

Eqs. (2.2) constitute a set of $6N$ coupled ordinary differential equations. It is well known from calculus that the integration of such a system amounts to finding $6N$ independent first integrals of motion. Hamiltonian systems constitute, however, a particular class in the sense that, under quite general conditions, it suffices to know only $3N$ first integrals. As Liouville demonstrated, if the latter are sufficiently regular mathematical functions of the variables, the system can be integrated by simple quadratures.

Liouville's theorem tells more. Under the above conditions it shows that it is, in principle, possible, by means of appropriate transformations which preserve the Hamiltonian structure and which are known as *canonical transformations* to cast eqs. (2.2) into the form (Goldstein, 1959; Lichtenberg and Lieberman, 1983)

$$\left. \begin{aligned} \mathrm{d}I_i/\mathrm{d}t &= 0 \\ \mathrm{d}\phi_i/\mathrm{d}t &= \omega_i(\{I_j\}) \quad i = 1, \ldots, 3N \end{aligned} \right\} \quad (2.3)$$

in which the I_is are suitable combinations of the constants of motion. We refer to I_i and ϕ_i as action and angle variables, respectively.

Eqs. (2.3) define the particular class of *integrable systems*. The behavior of such systems is easily obtained by straightforward integration:

$$\left. \begin{aligned} I_i &= I_{i_0} = \text{const.} \\ \phi_i &= \omega_i t + \phi_{i_0} \end{aligned} \right\} \quad (2.4)$$

If we look at (I_i, ϕ_i) as polar coordinates, in the system's original variables we realize that the behavior will be *multiperiodic*. If the frequencies $\{\omega_i\}$ are rational functions of each other, there will be a common divisor frequency and the behavior will actually be simply-periodic as alluded to already in Section 1.3. The general mathematical expression of this condition of *commensurability* is that there are integers $k_1, k_2, \ldots$, not all zero, such that

$$k_1\omega_1 + k_2\omega_2 + \ldots = 0 \qquad (2.5)$$

If there exists no such set $\{k_1, k_2, \ldots\}$ the frequencies will be irrationally related and the multiperiodic motion described by (2.3)–(2.4) will be *quasi-periodic*. Such an aperiodic motion may look quite intricate. Relation (2.5) defines the important concept of *resonance*, the relevance of which will become clear later on in this book (see especially Chapter 5).

Any Hamiltonian system with one degree of freedom, that is to say, one pair (q_1, p_1) of coordinates and momenta, is integrable since it possesses one regular constant of motion, the total energy H. The hoop (Section 1.2), the simple pendulum and the linear harmonic oscillator are therefore integrable systems. A Hamiltonian system with two degrees of freedom is integrable if there exists a sufficiently regular first integral independent of H. For three degrees of freedom three first integrals are needed. In the two-body problem in the presence of central forces, whose importance stems from its relation to the motion of celestial bodies, their existence follows from conservation of linear and angular momentum (in addition, of course, to that of energy). More generally, all systems that can be separated into uncoupled systems of one degree of freedom are integrable. This is the basis of the extensive literature on small vibrations around a position of equilibrium, of which solid state physics is a particularly important illustration. Certain problems of nonlinear vibrations can also be integrable, like the Toda lattice or the Korteweg–de Vries equation (Infeld and Rowlands, 1990).

Integrable systems dominated mechanics for almost three centuries. During this period it was thought that all physically relevant systems belonged to this class. As a result, multiperiodic behavior was considered to be the typical – and even the only possible – behavior of a mechanical system. Today the situation looks very different. Since the 1950s, thanks to the historic contributions of A. Kolmogorov, it is realized that many (and, in fact, in a certain mathematical sense 'most') of the naturally occurring systems are nonintegrable (Lichtenberg and Lieberman, 1983; Arnol'd, 1980). This has some momentous consequences, since by the loss of integrability multiperiodicity is no longer guaranteed. This leaves room for aperiodic unstable behavior and, in particular, for *Hamiltonian chaos*. There is ample evidence of the ubiquity of this phenomenon, to which we shall come back later. For now we want to point out two important areas in which chaos in classical mechanics is contributing to a radical change of perspective. The first is celestial mechanics, in connection with the celebrated three-body problem (Moser, 1973; Sussman and Wisdom, 1992; Laskar and Robutel, 1993). The second relates to the microscopic foundations of statistical mechanics, the main question here being the passage from microscopic, time-reversible behavior as described by eqs. (2.2) to the macroscopic, time-irreversible behavior as described by hydrodynamics or chemical kinetics. Hamiltonian chaos provides the missing link in that it shows how probabilistic concepts can naturally emerge from a system described by a well-defined, perfectly deterministic set of laws (Krylov, 1979; Bunimovitch and Sinai, 1980; Penrose, 1970, 1979; Prigogine, 1962, 1980).

Let us now illustrate the laws of classical mechanics using the example of the hoop considered in Section 1.2. We first recall that, as the hoop is a one degree of freedom system, it cannot give rise to chaotic behavior. Nevertheless, as we shall see in Chapters 4 and 5, it will give rise to the phenomenon of instability and bifurcation, in full agreement with the experimental results described in Section 1.2.

The starting point is eq. (2.1)

$$m \, dv_t/dt = F_t \tag{2.6}$$

v_t, F_t being the tangential velocity and force (Fig. 1.1). Writing

$$\left.\begin{aligned}
v_t &= r \, d\theta/dt \\
F_t &= (\text{weight} + \text{centrifugal force})_t \\
&= -mg \sin \theta + m\omega^2 r \sin \theta \cos \theta
\end{aligned}\right\} \tag{2.7}$$

we obtain from eq. (2.6)

$$\frac{d^2\theta}{dt^2} = \frac{g}{r}\sin\theta\,(\lambda\cos\theta - 1) \qquad (2.8a)$$

where we have set

$$\lambda = \omega^2 r/g \qquad (2.8b)$$

This equation exhibits quite clearly the nonlinearity inherent in the problem. It also shows nicely how the *constraint* enters the dynamics in a natural manner, here through the dimensionless parameter λ which expresses the relative importance of the two adverse factors – gravity and centrifugal force – present in this problem.

It is instructive to consider the vicinity of the threshold value ω_c around which the system switches to a new equilibrium position (Section 1.2). Since in this range θ is small (if initially small), one can expand eq. (2.8a) in powers of θ and keep the first nontrivial terms. The result is:

$$d^2\theta/dt = \mu\theta + \nu\theta^3 \qquad (2.9a)$$

with

$$\left.\begin{aligned} \mu &= \frac{g}{r}(\lambda - 1) \\[2mm] \nu &= \frac{g}{6r}(1 - 4\lambda) \end{aligned}\right\} \qquad (2.9b)$$

Eq. (2.9a), also known as the *Duffing oscillator*, presents a number of interesting features:

(i)   It remains invariant under the substitution $\theta' = -\theta$. This reflects the invariance of the original problem with respect to reflections around the vertical axis.

(ii) It is largely independent of the details of the original problem, which enter only through the specific values of the parameters μ and ν. Actually, as will become clear in Chapter 5, in the small θ limit eq. (2.9a) is universal, and describes any nonlinear oscillator enjoying the symmetry property under (i).

(iii) It reduces to the classical harmonic oscillator when the nonlinearity is neglected, provided that μ can be written as $\mu = -\omega_0^2$, where ω_0 is a (real-valued) angular frequency. According to the first relation of (2.9b) this can only be the case if $\lambda < 1$. Beyond the threshold value $\lambda_c = 1$, or $\omega_c = (g/r)^{1/2}$ the oscillatory character is lost and θ shows explosive behavior. Nonlinear terms then become necessary and are the ones that eventually are responsible for the bifurcation of the two new equilibria in this problem.

2.2 The macroscopic level: balance equation of a macrovariable

As we saw in Chapter 1, nonlinear behavior is also prominent at the macroscopic level. The state variables that are relevant at this level of description are the collective variables associated with statistical averages of microscopic quantities. Typical examples are bulk motion velocity, pressure, concentration of a chemical in a solvent, etc.

A variable $B(t)$ which refers to the system as a whole, like total energy or momentum, is called *extensive*. One can associate to it an intensive variable $b(\mathbf{r},t)$, describing local behavior, through

$$B(t) = \int_V d\mathbf{r}\, b(\mathbf{r}, t) \tag{2.10}$$

V being the volume occupied by the system. We are interested in how $B(t)$ evolves in time (Prigogine, 1947; De Groot and Mazur, 1962). As a rule, B will vary owing to two types of process (Fig. 2.1):

Exchanges with the external world, $d_e B/dt$. These can be modeled by a flux of b, $\mathbf{J}_b$ through the surface Σ surrounding V.
Internal processes, $d_i B/dt$. These are generated by the system itself even when it is completely isolated, and can be modeled as a rate of production of b per unit volume, σ_b.

It follows that

$$\frac{dB}{dt} = \frac{d_e B}{dt} + \frac{d_i B}{dt} = - \int_\Sigma d\,\Sigma \mathbf{J}_b \cdot \mathbf{n} + \int_V d\mathbf{r}\, \sigma_b \tag{2.11}$$

where $\mathbf{n}$ is the outward normal to the surface Σ. For a fixed but otherwise arbitrary Σ, applying Gauss' divergence theorem to the surface integral in

Fig. 2.1 Schematic representation of a system in a volume V separated from the environment by a surface Σ. $\mathbf{J}_b$ stands for the flux of the quantity B from the environment into the system (in units of B per unit of surface and time) and σ_b for the spontaneous production of B inside the system (in units of B per unit volume and time).

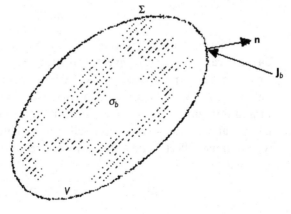

(2.11) and substituting B from (2.10) one obtains straightforwardly from the above the general balance equation for an intensive variable

$$\partial b(\mathbf{r}, t)/\partial t = -\operatorname{div} \mathbf{J}_b(\mathbf{r}, t) + \sigma_b(\mathbf{r}, t) \tag{2.12}$$

This equation is purely formal: since $\mathbf{J}_b$ and σ_b are unknown at this stage, it cannot provide as such information on the time development of b. Still, it is an interesting relation as it allows one to classify physical quantities into two large categories:

Conserved quantities, for which the source term $\sigma_b = 0$. Such quantities vary only through the fluxes exchanged with the external world and remain therefore constant in an isolated system. Typical examples are the mass or the momentum and energy in the absence of external forces. *Nonconserved* quantities, for which $\sigma_b \neq 0$. A typical example is the mass or the mole number of a chemical constituent in the presence of chemical reactions. In general σ_b has no definite sign. A notable exception is the entropy source term σ_s, which is bound to be a nonnegative quantity by the second law of thermodynamics.

In what follows we consider, in turn, the particular form of balance equations for these two types of quantities. As we shall see they will lead us quite naturally to the fields of fluid dynamics and chemical kinetics.

2.3 Conserved variables in a one-component system and the equations of fluid dynamics

Consider a one-component fluid (Landau and Lifshitz, 1959a; Guyon *et al.*, 1991). The most obvious example of a conserved variable for such a system is the mass density, $b = \rho$. Mass is transported only through the bulk motion, whose velocity we denote by $\mathbf{v}(\mathbf{r}, t)$. Setting therefore $\mathbf{J}_\rho = \rho\mathbf{v}$, $\sigma_\rho = 0$ in eq. (2.12) we obtain

$$\partial\rho/\partial t = -\operatorname{div} \rho\mathbf{v} \tag{2.13}$$

Consider next the momentum density, $\mathbf{b} = \rho\mathbf{v}$. Source terms for this quantity can only come from external forces, $\sigma_b = \rho\mathbf{f}$. As regards the flux we first have, as in the case of mass balance, the transport of $\rho\mathbf{v}$ with the bulk velocity giving the contribution $\rho\mathbf{vv}$, which is now a tensor of rank 2. In addition, however, momentum can be transported through thermal motion and interactions between adjacent volume elements of the fluid whose global effect is described by the pressure tensor, $\mathbf{P}$. Eq. (2.12) becomes therefore

$$\partial\rho\mathbf{v}/\partial t = -\operatorname{div}(\rho\mathbf{vv} + \mathbf{P}) + \rho\mathbf{f} \tag{2.14a}$$

Performing the time and space derivations and utilizing (2.13) one may further transform (2.14a) into

$$\frac{d\mathbf{v}}{dt} = \frac{\partial \mathbf{v}}{\partial t} + (\mathbf{v} \cdot \nabla)\mathbf{v} = -\frac{1}{\rho} \operatorname{div} \mathbf{P} + \mathbf{f} \qquad (2.14b)$$

where the *hydrodynamic derivative* d/dt expresses the total variation of $\mathbf{v}$ along the motion of the fluid.

Eqs. (2.14) feature ρ and $\mathbf{v}$ which are among the system's state variables, the external force term $\mathbf{f}$ which is supposed to be known, and the pressure tensor $\mathbf{P}$. As long as the last of these is not expressed in terms of the state variables we will not have at our disposal a closed set of equations from which the state variables can be evaluated. In fluid dynamics this closure is achieved by a series of *constitutive*, or *phenomenological relations* (De Groot and Mazur, 1962). Specifically, we decompose $\mathbf{P}$ into an ideal fluid part and a part accounting for dissipation,

$$\mathbf{P} = p\mathbf{1} + \mathbf{\sigma} \qquad (2.15)$$

where in the ideal part p is the hydrostatic pressure and in the dissipative part $\mathbf{\sigma}$ is the stress tensor expressing the frictional forces exerted between adjacent fluid regions. We adopt the *local equilibrium assumption*, whereby locally the thermodynamic quantities depend on the same variables as in equilibrium (De Groot and Mazur, 1962). It is shown in statistical mechanics that this assumption is reasonable as long as the constraints driving the system out of equilibrium vary in space and time on a scale that is much larger than the scales associated with molecular level processes. Under these conditions one may then, in an isotropic system,

(a) use for p the equation of state

$$p = p(\rho, T) \qquad (2.16a)$$

where T is the temperature;

(b) express $\mathbf{\sigma}$ as

$$\sigma_{ij} = -\eta \left(\frac{\partial v_i}{\partial r_j} + \frac{\partial v_j}{\partial r_i} - \frac{2}{3} \operatorname{div} \mathbf{v} \delta_{ij}^{kr} \right) - \zeta \operatorname{div} \mathbf{v} \delta_{ij}^{kr} \qquad (2.16b)$$

where η and ζ are, respectively, the shear and bulk viscosity coefficients. In general these phenomenological coefficients are T- and ρ-dependent but in most applications considered in this book they will be treated as constants.

Eqs. (2.14)–(2.16) now constitute a closed set provided that an evolution equation for the temperature T can be derived. We obtain this equation by considering the internal energy density ρu, the third important quantity characterizing the state of a fluid. We first recall that in the presence of conservative forces the total energy density ρe,

$$\rho e = \rho(\tfrac{1}{2}v^2 + u + \psi) \tag{2.17}$$

where ψ is the potential energy associated with the external forces, is conserved. From eq. (2.12) one has therefore

$$\partial \rho e / \partial t = -\operatorname{div} \mathbf{J}_e \tag{2.18}$$

In addition to the universal transport mechanism of ρe through the fluid velocity, $\rho e \mathbf{v}$, one must add a contribution associated with the mechanical work $\mathbf{P} \cdot \mathbf{v}$ performed by the internal stresses of the system as well as a purely dissipative contribution $\mathbf{J}_{th}$ arising from the transport of energy through thermal motion and intermolecular interactions:

$$\mathbf{J}_e = \rho\, e\mathbf{v} + \mathbf{P} \cdot \mathbf{v} + \mathbf{J}_{th} \tag{2.19}$$

Substituting into eq. (2.18) and taking into account eqs. (2.13)–(2.15) one arrives at an equation for the internal energy u,

$$\rho \frac{du}{dt} = \rho \left[\frac{\partial u}{\partial t} + (\mathbf{v} \cdot \nabla)u \right] = -\operatorname{div} \mathbf{J}_{th} - \mathbf{P} \cdot \operatorname{grad} \mathbf{v} \tag{2.20}$$

As in the case of momentum balance, we need constitutive relations to close this equation. In the local equilibrium regime these relations are:

(a) an equation of state

$$u = u(\rho, T) \tag{2.21a}$$

(b) Fourier's law of heat conduction

$$\mathbf{J}_{th} = -\lambda \nabla T \tag{2.21b}$$

where λ is the heat conductivity coefficient.

Eqs. (2.13), (2.14) and (2.20) supplemented with (2.15), (2.16) and (2.21) and appropriate boundary conditions constitute the closed set of equations of fluid dynamics from which the five hydrodynamic fields ρ, T and the three components of $\mathbf{v}$ can, in principle, be evaluated. On inspecting these equations one immediately realizes that they all feature a *universal nonlinearity* of the form $\mathbf{v} \cdot \nabla x$, x being ρ, $\mathbf{v}$ or T, reflecting the fact that the properties of a fluid are transported by the fluid velocity – itself a fluid property. We may refer to this mechanism as the 'hydrodynamic feedback'. It is worth noting that this nonlinearity subsists in the limit of ideal fluids in which all dissipative effects are neglected. On the other hand as one sees from (2.16b) and (2.21b) dissipative effects typically give rise to linear contributions, unless the phenomenological coefficients η, ζ and λ are state-dependent. The complex behaviors surveyed in Chapter 1 in connection with the Bénard problem stem, therefore, primarily from the 'inertial' nonlinearity inherent in fluid mechanics.

2.4 Nonconserved variables in a multicomponent system and the equations of chemical kinetics

We now consider a multicomponent system involving n chemically active constituents $i = 1, \ldots, n$ participating in r chemical reactions $\rho = 1, \ldots, r$ (Nicolis and Prigogine, 1977; Aris, 1975). In addition to the variables considered in the preceding section, in order to describe the macroscopic state of the system, we also need the composition variables $\{X_i(\mathbf{r}, t)\}$. These may be mass densities, molar densities or mole fractions but in what follows we will usually argue in terms of molar densities. According to eq. (2.12),

$$\partial X_i/\partial t = -\operatorname{div} \mathbf{J}_i + \sigma_i \qquad (2.22)$$

In addition to the universal transport mechanism of X_i through the bulk motion velocity, $X_i\mathbf{v}$, one must also consider a purely dissipative contribution $\mathbf{J}_i^{\mathrm{diff}}$ associated with diffusion, that is to say, transport through thermal motion and intermolecular interactions,

$$\mathbf{J}_i = X_i\mathbf{v} + \mathbf{J}_i^{\mathrm{diff}} \qquad (2.23a)$$

In the local equilibrium regime the diffusion flux is given by Fick's law,

$$\mathbf{J}_i^{\mathrm{diff}} = -D_i\nabla X_i \qquad (2.23b)$$

in which the additional simplifying assumption of an ideal mixture has been made to allow us to neglect cross-diffusion terms.

Let us turn now to the source term σ_i in (2.22), which reflects the effect of chemical reactions. We write formally a chemical reaction as

$$\sum_{i=1}^{n} \bar{v}_{i\rho}X_i \overset{k_\rho(T)}{\rightarrow} \sum_{j=1}^{n} \bar{v}'_{j\rho}X_j \quad \rho = 1, \ldots, r \qquad (2.24)$$

in which $\bar{v}$, $\bar{v}'$ stand for the number of moles of the reactants and products involved in the reaction and k_ρ for the rate constant. The difference $v_{i\rho} = \bar{v}'_{i\rho} - \bar{v}_{i\rho}$ gives the number of moles produced ($v_{i\rho} > 0$) or consumed ($v_{i\rho} < 0$) in the process. Hence, if we denote by w_ρ the velocity of the reaction we are entitled to write σ_i as

$$\sigma_i = \sum_\rho v_{i\rho} w_\rho \qquad (2.25a)$$

Again, we need a phenomenological closure relation to express w_ρ in terms of the state variables. In ideal systems this is provided by the *law of mass action*,

$$w_\rho = k_\rho(T) \prod_j X_j^{\bar{v}_{j\rho}} \qquad (2.25b)$$

which expresses that the rate of reaction ρ is proportional to the frequency of encounters between the molecules of the chemicals participating in the reaction times a coefficient k_ρ, which is generally state-dependent, specifying the fraction of those encounters that will actually result in a change of chemical identity. In an ideal system the frequency of encounters is simply the product of concentrations, but in a nonideal one the effect of intermolecular interactions must be taken into account through the activity coefficients.

Thanks to eqs. (2.23) and (2.25) the set of equations for X_i combined with the equations of fluid dynamics of Section 2.3 is now closed. Let us look in detail at the structure of these equations in some important limiting cases.

A Well-stirred open reactors

Such reactors have already been introduced in Section 1.4, Fig. 1.9. Stirring inside the reactor entails that $\mathbf{J}_i^{\text{diff}} = 0$ throughout. The remaining flux term, $- \operatorname{div} X_i \mathbf{v}$ is nonvanishing only at the reactor's boundaries. In the part communicating with the feed streams it describes the entry of fresh products at a concentration X_i^0 and in the outlet it describes the evacuation process at the reactant concentration X_i inside the reactor. Introducing the residence time τ (eq. 1.2), we may therefore write the balance equations in a well-stirred open reactor in the form

$$\frac{dX_i}{dt} = \sum_\rho v_{i\rho} k_\rho \prod_j X_j^{\bar{v}_{j\rho}} + \frac{1}{\tau_i}(X_i^0 - X_i) \qquad (2.26)$$

B Reaction–diffusion equations

We consider next an unstirred reactor in mechanical equilibrium, $\mathbf{v} = 0$, and at constant temperature. Furthermore we take the diffusion coefficients D_i to be constant, an assumption that is reasonable as long as the system is not close to a phase instability leading to unmixing. Substituting

(2.23b) and (2.25b) into eq. (2.22) one then obtains the *reaction – diffusion equations*.

$$\partial X_i/\partial t = \sum_\rho v_{i\rho} k_\rho \prod_j X_j^{\bar{v}_{j\rho}} + D_i \nabla^2 X_i \qquad (2.27)$$

These partial differential equations, supplemented with appropriate boundary conditions which express the way the reactor is fed from the outside world, constitute a closed set of *n* partial differential equations since they decouple completely from the equations of fluid dynamics.

Eqs. (2.26) and (2.27) are *intrinsically nonlinear*, since a collision process is bound to involve the product of at least two concentrations. In contrast with hydrodynamic nonlinearity, this chemical nonlinearity arises through the dissipative terms in the balance equation. Furthermore it is system-specific since it involves explicitly the mole numbers $\bar{v}_{j\rho}$ and therefore, ultimately, the nature of the reaction mechanism. On the other hand, being independent of transport, chemical nonlinearities subsist even in the limit of a spatially uniform system (cf. eqs. (2.26)). Chemical kinetics provides, therefore, one of the very few genuine examples of nonlinear dissipative systems whose dynamics, although possibly very complex, is governed by a finite and sometimes very small number of degrees of freedom. In contrast to this, hydrodynamic nonlinearity is invariably associated with spatial inhomogeneities. As a result in a typical fluid dynamics problem one is confronted, at the outset, with a system involving an infinite number of degrees of freedom.

2.5 The Bénard problem: quantitative formulation

As a concrete illustration of the equations of evolution of macroscopic observables we outline in this section the quantitative formulation of the problem of thermal convection described in Section 1.3 (Chandrasekhar, 1961; Manneville, 1991). In the idealized situation of a horizontal shallow layer one expects the density variations along the fluid motion to be negligible compared to the density itself or, in other words, the logarithmic hydrodynamic derivative $(1/\rho)(\mathrm{d}\rho/\mathrm{d}t)$ to be vanishingly small. Comparing with the mass balance equation (2.13) we deduce that

$$\mathrm{div}\,\mathbf{v} = 0 \qquad (2.28)$$

which will be hereafter referred to as the *incompressibility condition*.

Let us turn to the momentum balance, eq. (2.14). In a closed cell surface effects can be ignored and the only force present is gravity,

$$\mathbf{f} = -g\mathbf{1}_z \qquad (2.29)$$

1_z being the (upward directed) unit vector along the vertical. Introducing the phenomenological relations (2.15) and (2.16), assuming that the phenomenological coefficients are state-independent and taking the incompressibility condition (2.28) into account we arrive at the *Navier–Stokes equation*

$$\rho\left[\frac{\partial \mathbf{v}}{\partial t} + (\mathbf{v}\cdot\nabla)\mathbf{v}\right] = -\nabla p - \rho g \mathbf{1}_z + \eta\nabla^2\mathbf{v} \qquad (2.30)$$

According to the arguments at the beginning of this section, in a shallow layer of fluid one might by tempted to treat ρ as a constant. However, if this property is applied everywhere in eq. (2.30) one observes that momentum balance becomes completely uncoupled from the thermal constraint which, according to Section 1.3, should play an essential role in the onset of thermal convection. We must therefore somehow incorporate in eq. (2.30) the effect of density variation. Now, according to the equation of state (2.16a),

$$\rho = \rho(p, T) \qquad (2.31)$$

In this relation the effect of pressure can be neglected owing to the shallowness of the layer (isothermal incompressibility). Furthermore, as long as temperature variations remain moderate one can expand (2.31) around a fixed reference state (ρ_0, T_0), hereafter chosen to correspond to the state at the lower plate, and write

$$\rho \approx \rho_0[1 - \alpha(T - T_0)] \qquad (2.32)$$

where α is the coefficient of thermal expansion. Typical values of this coefficient are $\alpha \sim 10^{-4}\,\mathrm{K}^{-1}$ for water and $\alpha \sim 10^{-3}\,\mathrm{K}^{-1}$ for air, justifying *a posteriori* the reasonableness of the expansion for temperature differences of the order of several degrees or so.

It follows from the above that the density variations in the fluid, as a function of temperature, are much smaller than the density itself, analogously for density variations as a function of pressure. This suggests that one may neglect the variation of ρ in the left hand side of Navier–Stokes equation since, in a typical laboratory experiment, the velocities and accelerations involved are small and vary smoothly in space. We cannot apply a similar simplification to the right hand side since, as observed earlier, this would eliminate the physical mechanism at the very basis of thermal convection. We thus arrive at the following 'minimal' version of Navier–Stokes equation,

$$\rho_0\left[\frac{\partial \mathbf{v}}{\partial t} + (\mathbf{v}\cdot\nabla)\mathbf{v}\right] = -\nabla p - \rho g \mathbf{1}_z + \eta\nabla^2\mathbf{v} \qquad (2.33)$$

Let us turn finally to the internal energy balance, eq. (2.20). Decomposing **P** as in (2.15) we write the internal stress term **P** · grad **v** as

$$\mathbf{P} \cdot \text{grad } \mathbf{v} = p \text{ div } \mathbf{v} + \boldsymbol{\sigma} \cdot \text{grad } \mathbf{v}$$

The first part is zero on account of the incompressibility condition, eq. (2.28). In view of eq. (2.16b) the second part is quadratic in the velocity gradient, a small quantity in a Bénard experiment under laboratory conditions, and can therefore be neglected. Furthermore, in accordance with the arguments leading to eq. (2.33) we neglect the variation of density in the left hand side of (2.20) and replace the phenomenological coefficient λ in Fourier's law (2.21b) by a constant. It remains to express the internal energy u in terms of the experimentally more relevant variables ρ and T by the equation of state (2.21a) or, in the case of incompressible fluid, by the simplified form

$$\left. \begin{array}{l} u = u(T) \\ du = c_p \, dT \end{array} \right\} \quad (2.34)$$

(notice that $c_p = c_v$ in an incompressible fluid). Inserting all these expressions and simplifications into eq. (2.20) we obtain

$$\rho_0 c_p \left[\frac{\partial T}{\partial t} + (\mathbf{v} \cdot \nabla)T \right] = \lambda \nabla^2 T \quad (2.35)$$

Eqs. (2.28), (2.32), (2.33) and (2.35) are the fundamental equations governing thermal convection. They will be taken up again in Chapter 6 and will serve as one of our favorite models on which the concepts and tools of nonlinear science will be illustrated. We recall here that their specific form stems from: (*a*) the particular way of treating density variations by replacing ρ by a constant everywhere except in the gravity term; (*b*) neglecting the viscous heating term $\boldsymbol{\sigma} \cdot \text{grad } \mathbf{v}$; and (*c*) supposing that the phenomenological coefficients η, λ, α, c_p are state-independent. We refer to these assumptions as the *Boussinesq approximation*. We have justified this approximation here on the basis of intuitive arguments. A more systematic derivation exploiting the existence of a number of smallness parameters in the problem has been elaborated by De Boer (1986) and by Velarde and Gordon (1976). Notice that the Boussinesq approximation fails when the depth of the layer becomes appreciable. This is the case in many important real-world situations such as convection in the atmosphere or in the oceans.

A set of partial differential equations like (2.33) and (2.35) together with conditions (2.28) and (2.32) does not constitute a well-posed problem unless it is supplemented by appropriate *boundary conditions*. Since the fluid layer is supposed to extend indefinitely in the horizontal direction the only requirement is that T and $\mathbf{v}$ remain bounded as the coordinates (x, y) tend to $\pm \infty$. We may also model the infinite extent of the system along this plane by periodic boundary conditions. The situation is somewhat more involved along the z direction. There are, of course, two pairs of boundary conditions that impose themselves by the very definition of the Bénard problem. The first expresses that the plates are maintained at fixed temperatures,

$$T(x, y, 0) = T_0, \qquad T(x, y, d) = T_1 \tag{2.36}$$

The second expresses the fact that the fluid is confined, hence its velocity along the vertical direction must vanish on the plates

$$w(x, y, 0) = w(x, y, d) = 0 \tag{2.37}$$

where we have set $\mathbf{v} = (u, v, w)$. However, these conditions are insufficient since (2.33) is equivalent to an equation of sixth order in one of the velocity components. One needs, therefore, four more boundary conditions for the velocity field. As it turns out these conditions depend on the nature of the experimental setup. Two typical cases can be envisaged.

A Rigid boundaries

Suppose that the fluid is placed between two rigid plates. By continuity its velocity must then vanish identically on these plates or, in view of (2.37),

$$\left.\begin{array}{l} u(x, y, 0) = u(x, y, d) = 0 \\ v(x, y, 0) = v(x, y, d) = 0 \end{array}\right\} \tag{2.38}$$

It follows from (2.38) that the partial derivatives of u and v along any direction on a horizontal plane vanish identically. In particular,

$$\partial u / \partial x = 0, \partial v / \partial y = 0 \quad \text{in} \quad z = 0, d$$

Combining with the incompressibility condition (2.28), we may transform these conditions to a condition on the vertical component of the velocity,

$$(\partial w / \partial z)_{x, y, 0} = (\partial w / \partial z)_{x, y, d} = 0 \tag{2.39}$$

B Free boundaries

As we shall see in Chapter 6 the solution of the Bénard problem with the boundary conditions (2.37) and (2.39) is technically very involved. For the sake of mathematical simplicity we consider here the case of a fluid confined by free boundaries, that is to say, boundaries on which no viscous stress is exerted. Using the expression (2.16b) of the stress tensor along with the incompressibility condition (2.28) we express this property as

$$\sigma_{xz} = \sigma_{yz} = 0 \quad \text{in} \quad z = 0, d$$

or more explicitly,

$$\left. \begin{array}{l} \dfrac{\partial u}{\partial z} + \dfrac{\partial w}{\partial x} = 0 \\[2mm] \dfrac{\partial v}{\partial z} + \dfrac{\partial w}{\partial y} = 0 \end{array} \right\} \text{in} \quad z = 0, d \qquad (2.40)$$

In practice these conditions are difficult to realize except under micro-gravity conditions. On the other hand, in a fluid subjected to such conditions one of the mechanisms at the origin of the thermal convection instability – the gravity-induced Archimedes force – disappears. A more elaborate study shows that an instability can still occur owing to the temperature dependence of the *surface tension*, measuring the decrease of surface energy arising from the decrease of cohesion in the fluid as the temperature increases. We refer to this mechanism as the *Marangoni effect* (Guyon *et al.*, 1991).

As before one may transform (2.40) into conditions for the vertical component *w*. First, it follows from (2.37) that the derivatives of *w* along any horizontal direction vanish identically,

$$\partial w / \partial x = \partial w / \partial y = 0 \quad \text{in} \quad z = 0, d$$

Combining with (2.40) one deduces that

$$\partial u / \partial z = \partial v / \partial z = 0 \quad \text{in} \quad z = 0, d \qquad (2.41)$$

Consider now the incompressibility condition (2.28) differentiated once with respect to z:

$$\frac{\partial^2 u}{\partial z \partial x} + \frac{\partial^2 v}{\partial z \partial y} + \frac{\partial^2 w}{\partial z^2} = 0$$

Requiring that u, v are differentiable functions which possess bounded derivatives one can exchange the order of derivations in the first two terms. Taking (2.41) into account one obtains

$$(\partial^2 w / \partial z^2)_{x, y, 0} = (\partial^2 w / \partial z^2)_{x, y, d} = 0 \qquad (2.42)$$

A number of variants of the Bénard experiment can also be formulated along similar lines such as a rigid lower and a free upper boundary. This actually models the experiments performed by Bénard himself, in which the upper surface of the fluid was in contact with ambient air.

2.6 Some representative chemical models giving rise to nonlinear behavior

We have already stressed the fact that the form of the equations of chemical kinetics is system-specific. In chemistry, therefore, there is no straight analog of the Boussinesq equations of the previous section describing, say, a chemical oscillator or a chemical system giving rise to pattern formation whatever its specific characteristics might be. In this section we survey a number of models that have been proposed over the years to interpret various types of complex nonlinearity-driven behavior. Many of these will be taken up in the subsequent chapters and used to illustrate the concepts and tools of nonlinear science.

We start once again with the BZ reaction. The first successful model of oscillatory behavior in this system has been developed by Noyes and coworkers on the basis of the mechanism discussed in Section 1.4. It is frequently referred to in the literature as the 'Oregonator'.

Let $X = [HBrO_2]$, $Y = [Br^-]$, $Z = 2[Ce^{4+}]$ be the concentrations of the three key substances featured in the mechanism of Section 1.4. From the analysis of this section we see that reaction ($A1$) describes the conversion of Y to X, reaction ($A3$) the simultaneous inactivation of X and Y, reactions ($B2$) and ($B3$) the autocatalytic generation of X, reaction ($B1$) the bimolecular decomposition of X, and the global reaction (C) the regeneration of Y from Z. Hence, we write the following steps (Field et al., 1972):

$$
\left.
\begin{aligned}
A + Y &\xrightarrow{k_1} X \\[4pt]
X + Y &\xrightarrow{k_2} P \\[4pt]
B + X &\xrightarrow{k_3} 2X + Z \\[4pt]
2X &\xrightarrow{k_4} Q \\[4pt]
Z &\xrightarrow{k_5} fY
\end{aligned}
\right\} \quad (2.43)
$$

Here the concentrations $A = B = [BrO_3^-]$ are supposed to remain fixed ('pool' chemical approximation (Gray, 1990)). P and Q denote waste products, and f a suitable stoichiometric coefficient. Note that all reactions are taken to be irreversible (the reversible version of this model has been studied by Field (1975)). The rate constants k_i contain the effect of H^+, of bromomalonic acid and any other species considered to act as a 'reservoir'. Numerical values of these constants can be inferred by comparison with the detailed mechanism. They turn out to be widely different, ranging from 1 to 10^9 M^{-1} s^{-1}. This entails that the system evolves according to different time scales, a property that gives rise to sharp, relaxation-like oscillations even close to the transition threshold to oscillatory behavior. This is confirmed by experiment.

Historically, the first chemical model giving rise to sustained oscillations while being fully compatible with the fundamental laws of physics and chemistry was a two-variable autocatalytic model known as the Brusselator (Prigogine and Lefever, 1968; Lefever and Nicolis, 1971; Lefever, Nicolis and Borckmans, 1988). Here the unique source of nonlinearity is the autocatalytic synthesis of X according to

$$
\left.
\begin{aligned}
&A \xrightarrow{k_1} X \\[4pt]
&B + X \xrightarrow{k_2} Y + C \\[4pt]
&2X + Y \xrightarrow{k_3} 3X \\[4pt]
&X \xrightarrow{k_4} D
\end{aligned}
\right\} \quad (2.44)
$$

in which the initial product concentrations A and B are again treated as fixed parameters and C, D denote waste products. In the presence of diffusion the Brusselator also generates a variety of spatial patterns (Nicolis and Auchmuty, 1974; Nicolis and Prigogine, 1977), including inhomogeneous stationary states similar to those discovered recently in the experiments of the Bordeaux and Austin groups (cf. Section 1.4) and spatio-temporal chaos (Kuramoto, 1984). This establishes the possibility of complex nonlinear behavior in purely dissipative systems, even in the absence of the inertial effects inherent in fluid dynamics.

Chemical reaction models involving a single variable may also give rise to highly nontrivial behavior in the form of multiple stationary states and propagating wave fronts. Such phenomena have been observed experimentally, notably in the iodate–arsenous acid reaction (Ganapathisubramanian and Showalter, 1983) and, even more typically, in combustion.

An elegant prototype mechanism for this type of behavior in isothermal system is provided by Schlögl's first and second models (Schlögl 1971, 1972).

$$
\left.
\begin{aligned}
&A + X \underset{k_2}{\overset{k_1}{\rightleftharpoons}} 2X \\[6pt]
&X \underset{k_4}{\overset{k_3}{\rightleftharpoons}} B
\end{aligned}
\right\} \quad (2.45a)
$$

and

$$
\left.
\begin{array}{c}
A + 2X \underset{k_2}{\overset{k_1}{\rightleftharpoons}} 3X \\[2mm]
X \underset{k_4}{\overset{k_3}{\rightleftharpoons}} B
\end{array}
\right\} \quad (2.45b)
$$

in which X denotes the unique variable and A, B are again treated as parameters. As in the Brusselator, the unique source of nonlinearity is the autocatalytic production of X.

A minimal one-variable model incorporating thermal effects is the Semenov–Frank-Kamenetskii model (Frank-Kamenetskii, 1969; Zeldovich, Barenblatt, Librovich and Makhviladze, 1985). Consider a well-stirred reactor in mechanical equilibrium, closed to mass transfer but capable of exchanging energy with a thermal reservoir at constant temperature T_0. The chemical transformation taking place within the reactor is an irreversible, unimolecular exothermic decomposition of a fuel in gas phase (cf. eq. (1.3)). For simplicity one assumes that the concentration of the reactant varies on a scale that is much slower than heat transfer and may thus be taken as a constant (equal to its initial value c_0). The only remaining relevant variable is therefore the temperature T. To obtain an evolution equation for this variable we turn to the general energy balance equation (eq. (2.20)). We set $\mathbf{v} = 0$ (mechanical equilibrium) and model the heat transfer term in a way analogous to the mass transfer term in the open well-stirred reactor (eq. (2.26)), known as Newton's cooling law,

$$
\frac{1}{V} \int_{\text{reactor}} d\mathbf{r}(-\operatorname{div} \mathbf{J}_{\text{th}}) = \alpha(T_0 - T) \tag{2.46a}
$$

In the presence of chemical transformations the equation of state (eq. (2.21a)) linking internal energy to T also features the degree of advancement ξ of the reaction, defined by $d\xi/dt$ = reaction velocity w. It follows that

$$
\begin{aligned}
\rho \frac{du(T, \xi)}{dt} &= \rho \left(\frac{\partial u}{\partial T}\right)_\xi \frac{dT}{dt} + \rho \left(\frac{\partial u}{\partial \xi}\right)_T \frac{d\xi}{dt} \\[2mm]
&= \rho c_v \frac{dT}{dt} + \Delta H w \\[2mm]
&= \rho c_v \frac{dT}{dt} + k c_0 \Delta H
\end{aligned} \tag{2.46b}
$$

where ρ is the mass density of the mixture, c_v the specific heat at constant volume and $-\Delta H$ the heat of reaction. Substituting (2.46a) and (2.46b) into eq. (2.20) and taking into account the Arrhenius law for the temperature dependence of the rate constant k (eq. (1.4)) we arrive at the Semenov equation

$$\rho c_v \frac{dT}{dt} = -k_0 c_0 \Delta H \, e^{-E/RT} + \alpha(T_0 - T) \qquad (2.47)$$

As it turns out this equation quite successfully describes the first stages of thermal explosion characteristic of combustion phenomena. When supplemented with (heat) diffusion terms to model a nonstirred reactor (Frank-Kamenetskii equation) it also gives rise to propagating fronts thereby providing the basis for understanding the formation of a flame.

As stressed in Section 1.5, in many cases the kinetics of biologically relevant phenomena is completely isomorphic to the kinetics of chemical reactions. An interesting illustration of this analogy is provided by population dynamics. Let X be the density of the population (number of individuals per surface) in the supporting medium and A the density of resources. One may reasonably decompose the processes contributing to the evolution of X as follows:

Birth,

$$A + mX \xrightarrow{k_1} nX \quad (n > m) \qquad (2.48a)$$

Death,

$$X \xrightarrow{k_2} D \qquad (2.48b)$$

Migration,

$$B \xrightarrow{k_3} X \qquad (2.48c)$$

Regulation, consisting of the slowing down of the rate of growth k_1 as the density of individuals increases and it becomes difficult for the supporting medium to sustain further growth. A minimal model for this phenomenon is to set

$$k_1(X) = a - bX \qquad (2.48d)$$

This law was first stipulated by the Belgian scientist Verhulst (Verhulst, 1845). At this time it was in complete opposition to the prevailing theories of Malthusian growth (exponential growth or decay, according to whether the birth rate is larger or smaller than the death rate). Notice the striking resemblance of (2.48a)–(2.48d) with the four steps of Schlögl's first model, eq. (2.45a).

From (2.48a)–(2.48d) neglecting for simplicity migration and setting $m = 1$, $n = 2$ we arrive at the rate equation

$$\mathrm{d}X/\mathrm{d}t = (a - bX)AX - k_2 X \qquad (2.49)$$

or, setting

$$k = aA - k_2, \quad N^{-1} = \frac{bA}{aA - k_2} \qquad (2.50)$$

in the more compact form

$$\frac{\mathrm{d}X}{\mathrm{d}t} = kX\left(1 - \frac{X}{N}\right) \qquad (2.51)$$

This is the celebrated *logistic equation*. It describes in a surprisingly successful manner population growth on various scales (Montroll and Badger, 1974) as a process in which after an initial exponential growth stage (small X, $k > 0$) the system performs a fast transition toward a plateau value $X = N$, N being the ecosystem capacity (Fig. 2.2). It should be stressed that in the real world N is not constant but evolves in time on a slow scale as a result of evolution or technological innovation. After having reached a first plateau (representative, say, of an agricultural society) X may thus subsequently jump to a number of successively higher plateaus reflecting, for instance, the concentration of individuals in urban centers.

The above reasoning can easily be extended to include coexisting

Fig. 2.2 Time dependence of the solution of the logistic equation, eq. (2.51). Parameter values: $N = 1$, $k = 0.1$. Initial condition: $X_0 = 0.01$.

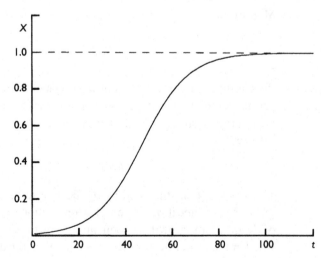

populations or predator – prey interactions. It should be noted that in population biology the generation time is often comparable to the time scale of the dynamics (May, 1974). This entails that the continuous time logistic equation must sometimes be replaced by its discrete counterpart

$$x_{n+1} = ax_n - bx_n^2 \qquad (2.52)$$

which, as will be seen in Chapter 7, may give rise to a surprising variety of complex behaviors.

Problems

2.1 Construct the explicit form of the transformation to action-angle variables mapping the harmonic oscillator hamiltonian, $H = p^2/2m + (k/2)q^2$ into the form $H = \omega I$. Identify the value of ω in terms of the original parameters.

2.2 Write out the equation generalizing (2.8a) in the presence of friction assuming that the friction force is a constant multiple of the velocity of the mass. Derive the equation replacing (2.9a) in this case and comment on its symmetry properties with respect to the transformation $\theta \to -\theta$ and $t \to -t$.

2.3 Derive in the Bénard problem a quantitative expression for the Archimedes force and for the viscous frictional force acting on a small spherical fluid volume of radius R displaced upward by a length element δz by a random disturbance. Comment on the onset of the instability by analyzing the relative magnitude of these two forces.

2.4 Carry out the detailed derivation of eq. (2.20) starting from eqs. (2.13)–(2.19).

2.5 Derive the entropy balance equation for the Bénard problem in the Boussinesq approximation using the local equilibrium assumption (Section 2.3). Hint: express formally the entropy density as a function of u and ρ and use the balance equations for these quantities (De Groot and Mazur, 1962).

2.6 Derive the balance equations of the Brusselator, eq. (2.44) in a CSTR, assumed closed to X, Y and open to the transfer of A, B, C, D with $X_A^\circ \neq 0$, $X_B^\circ \neq 0$, $X_C^\circ = X_D^\circ = 0$ and $\tau_A = \tau_B = \tau_C = \tau_D = \tau$.

2.7 The pool chemical approximation (Section 2.6) amounts to neglecting in a chemical reaction the consumption of certain initial products during the time scale of variation of the reaction intermediates (Gray, 1990). Derive the rate equations for A, B, X, Y in the Brusselator (eq. (2.44)) viewed as a closed system. Discuss in terms of the values of the rate constants k_i limiting cases where the equations reduce to a pair of equations for X and Y corresponding precisely to the pool chemical approximation (Lefever *et al.*, 1988).

2.8 Identify the nondissipative, time reversible and the dissipative, time irreversible contributions in the evolution equations of the Bénard problem and of a reaction–diffusion system. Check the invariance of these equations with respect to space and time translations.

2.9 In Section 2.4 a distinction was made between 'kinetic' and 'thermodynamic' nonlinearities, the former being already present in the nondissipative limit and the second arising from the constitutive relations linking the fluxes with the constraints. Consider the isomerization reaction $A \underset{k_2}{\overset{k_1}{\rightleftharpoons}} B$. As shown in thermodynamics (De Groot and Mazur, 1962) the driving force of this reaction, also known as *affinity* is $\mathscr{A} = \mu_A - \mu_B$, μ being the chemical potential. Assuming an ideal mixture, relate the flux $J = w_1 - w_2$ to this driving force. Discuss in terms of the distance from the state of chemical equilibrium $(w_{1,eq} = w_{2,eq})$ limiting cases reducing this thermodynamically nonlinear law to a linear one.

2.10 In the absence of external forces the general relation linking the diffusion flux in an n-component mixture to the corresponding constraint is (De Groot and Mazur, 1962)

$$J_i^{\text{diff}} = -\sum_{j=1}^{n-1} L_{ij} \nabla \left(\frac{\mu_j - \mu_s}{T} \right)$$

where L_{ij} is a set of phenomenological coefficients and the nth component s plays the role of the solvent. Show how in an ideal mixture involving one solute species in a solvent this law reduces to (2.23b) and relate Fick's coefficient D_i to the phenomenological coefficient L_{ii}.

2.11 Derive the extended form of eq. (2.47) to account for the consumption of the reactant. What is the stationary state solution of the new equations?

CHAPTER THREE

Dynamical systems with a finite number of degrees of freedom

3.1 General orientation

As we saw in Chapter 2 the evolution of the state variables of a system obeying the laws of classical physics is given by a set of differential equations of first order in time. These may be *ordinary* (ode) like Hamilton's equations (2.2) and the equations of chemical kinetics in a well-stirred reactor (eqs. (2.26)) or *partial* (pde), like the equations of fluid dynamics (eqs. (2.13), (2.14)–(2.16) and (2.20)–(2.21)) and the reaction–diffusion equations (eqs. (2.27)). In this latter case, which is typical in a macroscopic description, one deals in principle with an infinity of degrees of freedom – the values of the state variables (which are now fields) at each point in space as functions of time.

Although in certain problems this may constitute an essential aspect of the phenomenon under consideration, in other cases it may happen that this description can effectively be reduced to a finite number of variables. As an example consider the typical form of a set of partial differential equations.

$$\partial X_i/\partial t = F_i(\{X_j(\mathbf{r}, t)\}, \{\nabla^k X_j(\mathbf{r}, t)\}) \tag{3.1}$$

We endow our functional space with a scalar product and a norm and introduce a complete basis of orthonormal functions $\{\phi_m(\mathbf{r})\}$. Expanding $X_i(\mathbf{r},t)$ in the basis

$$X_i(\mathbf{r}, t) = \sum_m c_{im}(t)\phi_m(\mathbf{r}) \tag{3.2}$$

substituting into eq. (3.1) and using the orthogonality properties of the basis one easily arrives at

$$\frac{\mathrm{d}c_{im}(t)}{\mathrm{d}t} = \left(\phi_m, F_i\left(\left\{\sum_n c_{jn}(t)\phi_n(\mathbf{r})\right\}, \left\{\sum_n c_{jn}(t)\nabla^k\phi_n(\mathbf{r})\right\}\right)\right) \tag{3.3}$$

For any given form of evolution laws F_i the scalar product (integration over space) can be carried out explicitly in eq. (3.3), and one is left with an infinite set of coupled odes for the expansion coefficients c_{im}. The coupling reflects the fact that the *spatial modes* described by $\phi_m(\mathbf{r})$ do not evolve independently: even if initially only one mode is present $(c_{im} = c_{i\alpha}\delta_{m\alpha}^{kr})$, a *cascading process* will be switched on whereby all modes will sooner or later become excited. Nevertheless, it so happens that in many cases the amplitudes of these modes are scaled by a smallness parameter, thereby allowing the infinite hierarchy of eqs. (3.3) to be truncated to a finite order. Additional simplifications may arise by the judicious use of symmetry arguments.

One famous example of a reduction of the above type, usually referred to in the literature as the *Galerkin method* (Kantorovitch and Krylov, 1964), is the Lorenz equations (Lorenz, 1963). They arise by expanding the velocity and temperature fields in the Bénard problem in Fourier series and by truncating the Fourier amplitude equations deduced from (2.33) and (2.35) to three modes. The resulting system of odes reads

$$
\begin{aligned}
dX/dt &= \sigma(-X + Y) \\
dY/dt &= rX - Y - XZ \\
dZ/dt &= XY - bZ
\end{aligned}
\qquad (3.4)
$$

where X, Y, Z are rescaled Fourier amplitudes and σ, r, b positive combinations of the original parameters of the Bénard problem.

The present chapter deals with systems that are amenable to description in terms of a finite number of variables. Assuming, in addition, that the constraints acting on such systems do not depend explicitly on time (*autonomous* systems) we write the typical form of the evolution laws as

$$
dX_i/dt = F_i(\{X_j\}, \lambda) \quad i = 1, \ldots, n
\qquad (3.5)
$$

or, introducing the vector notation

$$
\mathbf{X} = \text{column}(X_1, \ldots, X_n)
$$

in the more compact form

$$
d\mathbf{X}/dt = \mathbf{F}(\mathbf{X}, \lambda)
\qquad (3.6)
$$

In writing eqs. (3.5) and (3.6) we have also accounted for the fact that a real-world system involves a number of parameters λ, hereafter referred to as *control parameters*, reflecting its internal structure (viscosity or diffusion coefficients) or the way it communicates with the external world (thermal or shear constraints, residence time of a chemical pumped into a reactor, etc). We know already from Chapter 1 that such parameters play an important role in the system's behavior, it is therefore natural to incorporate them into our description.

Even in the simplified form of eqs. (3.6) in which space dependencies are discarded, one is left with an intractable problem: the methods of modern science do not allow one to derive explicit solutions of eqs. (3.6) as soon as the number of variables n is larger than two, except in some pathological limiting cases. The main reason for this is the *nonlinearity* inherent in the equations which, as stressed in Chapter 2, is a universal property of the vast majority of natural systems. We shall cope with this fundamental limitation by giving up the idea of a full quantitative understanding, thereby focusing on the *qualitative* aspects of the dynamics. The basis of such a qualitative study is the notion of phase space.

3.2 Phase space

We embed the evolution of our system, as described by eqs. (3.5) and (3.6), into the abstract n-dimensional space spanned by the full set of variables $(X_1, \ldots, X_n)$, which we shall refer to from now on as the *phase space*, Γ. By definition, an instantaneous state of the system is given by a particular set of values of $(X_1, \ldots, X_n)$ – hence by a unique point P in phase space (Fig. 3.1). Conversely, a phase space point P can be characterized by its coordinates $(X_1, \ldots, X_n)$ and defines, therefore, a state of our system in a unique fashion. In other words, there is a one-to-one correspondence between physical states of the system under consideration and phase space points.

Consider now a succession of states $(X, \ldots, X_t, \ldots)$ attained in the course of time t. By the above argument this will determine in phase space a succession of points $(P, \ldots, P_t, \ldots)$ joined by a curve C, the phase

Fig. 3.1 Phase space trajectory of a dynamical system.

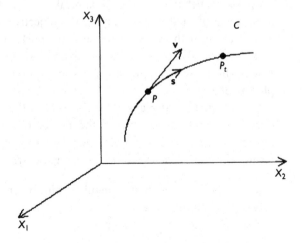

space trajectory (Fig. 3.1). Repeating the process for all possible histories $(X', \ldots, X'_t, \ldots)$ etc. one generates a continuous family of phase space trajectories, in other words, the evolution of a system amounts to a *mapping* of Γ into itself.

The tangent to the phase space trajectory at a given point is the phase space velocity $\mathbf{v} = (v_1, \ldots, v_n)$. By analogy with fluid mechanics it may be thought of as the velocity of a *flow* generated by the *vector field*, eqs. (3.6), in Γ. Its orientation angles relative to the axes are given by $\sigma_i = \mathrm{d}s/\mathrm{d}X_i$, s being the length element along the trajectory. Writing $\mathrm{d}s = (\sum_k \mathrm{d}X_k^2)^{1/2}$ and using eqs. (3.5) we obtain:

$$\sigma_i = \left[1 + \sum_{k \neq i} \left(\frac{\mathrm{d}X_k}{\mathrm{d}X_i} \right)^2 \right]^{1/2} = \left[1 + \sum_{k \neq i} \left(\frac{F_k}{F_i} \right)^2 \right]^{1/2} \qquad (3.7)$$

These quantities are well-defined everywhere in Γ except on points $\mathbf{X}_s = \{X_{js}\}$ given by the solution of the set of algebraic equations

$$F_1(\{X_{js}\}, \lambda) = \cdots = F_n(\{X_{js}\}, \lambda) = 0$$

We shall refer to these points as *singular points*. In an autonomous system the singular points remain fixed in phase space for all times – hence the terminology of *fixed points* also used to characterize them.

A system like in eqs. (3.6), endowed with an evolution law ϕ_t such that $\mathbf{X}(t) \equiv \mathbf{X}_t = \phi_t(\mathbf{X}_0)$ and embedded in Γ is referred to as a *dynamical system*. The set of smooth phase space trajectories (i.e. trajectories not containing fixed points) and singular points constitutes the *phase portrait* of the system. By virtue of the one-to-one correspondence between the succession of states in time and flow in phase space, one can assert that the determination of the phase portrait will give information on the full set of all possible behaviors of a dynamical system. In the more modest perspective of qualitative analysis, the objective will be limited to the classification of the types of phase space portrait that can be realized. Qualitative analysis is thus reduced to a well-defined geometric problem. It is for this reason that phase space is so central in the study of nonlinear phenomena.

One property that plays a decisive role in the structure of the phase portrait relates to the uniqueness theorem of the solutions of the ordinary differential equations underlying our dynamical system, eqs. (3.6). This result, which goes back to Cauchy (see e.g. Cesari (1963)), stipulates that:

if $\mathbf{X}_0$ is a point other than a singular point belonging to a certain open subset U of phase space Γ,
if $\mathbf{F}$ satisfies the Lipschitz property, i.e.

$$|\mathbf{F}(\mathbf{Y}) - \mathbf{F}(\mathbf{X})| \leqslant K\,|\,\mathbf{Y} - \mathbf{X}\,| \qquad (3.8)$$

for some $K < \infty$ and with $|\cdot|$ the Euclidean norm, then

there is an interval $t_0 < t < t_0 + T$ such that there exists in U a unique solution $\mathbf{X}(t;\mathbf{X}_0,t_0)$ satisfying (3.6) with the initial condition $\mathbf{X}(t_0) = \mathbf{X}_0$.

In the phase space representation, one immediately realizes that the theorem automatically rules out the intersection of two trajectories or the self-intersection of a given trajectory at any point other than a singular point. This introduces topological constraints delimiting the type of phase space motion. These constraints are particularly severe in one and two dimensions, and it is not an accident that complex dynamical behaviors in the form of deterministic chaos become possible only in three-or higher-dimensional continuous time dynamical systems.

3.3 Invariant manifolds

A second element of great importance in organizing the phase portrait of a dynamical system is the *compact invariant manifolds* that may exist in the flow. By this we mean objects embedded in the phase space that are bounded and are mapped onto themselves during the evolution generated by eqs. (3.6). By definition we exclude the trivial invariant manifold constituted by Γ itself, in other words, we limit ourselves to manifolds whose dimensionality d is strictly less than the phase space dimensionality n.

The *fixed points* encountered in the preceding section, eq. (3.8), are an obvious example of an invariant set, of dimension $d = 0$. Since by virtue of (3.6) $dX_i/dt = F_i = 0$ on these points, we conclude that fixed points in phase space describe the *stationary states* that can be reached by the underlying system, such as the state of the mechanical equilibrium in the example of the hoop (Section 1.2) or the stationary nonequilibrium values of chemical concentrations in an open reactor.

The next obvious example of a compact invariant manifold is given by one-dimensional objects in phase space in the form of closed curves free of fixed points (Fig. 3.2). Once on such a curve the system goes repeatedly through exactly the same states, in other words, it exhibits a periodic behavior. As we saw in Chapter 1 this type of behavior arises in large classes of natural systems. The importance of the geometrical view of dynamical systems is now beginning to be clear since it allows us to establish a correspondence between dynamical behavior and geometric figures embedded in phase space.

To proceed further it is instructive to investigate the nature of invariant

Fig. 3.2 A one-dimensional invariant manifold in the form of a closed curve C free of fixed points embedded in a three-dimensional phase space, coexisting with a zero-dimensional invariant manifold in the form of a fixed point P.

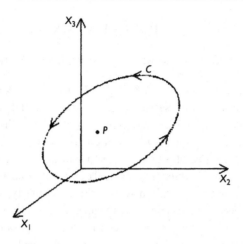

manifolds in conjunction with the dimensionality of the embedding space or, equivalently, the number of variables involved in the dynamics.

A One-variable systems

The phase space is one-dimensional. Since the dimensionality of the invariant manifold is strictly smaller than the phase space one, the only possibility one is left with are zero-dimensional manifolds – the fixed points. If we further require on physical grounds that the system remains bounded we arrive at a very restrictive picture of the motion in phase space, as illustrated in Fig. 3.3.

B Two-variable systems

Since now $n = 2$, d can be either 0 (fixed point) or 1 (invariant curve). A one-dimensional manifold can be a closed curve free of fixed points (Fig. 3.2). In a two-dimensional space the only way to realize this while avoiding at the same time self-intersection (in agreement with the uniqueness theorem) is through *one-circuit* closed curves. However, in view of the possible coexistence with fixed points, more intricate

Fig. 3.3 Phase portraits of a one-dimensional dynamical system: (a) a single fixed point; (b) several coexisting fixed points.

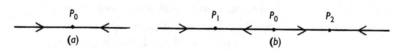

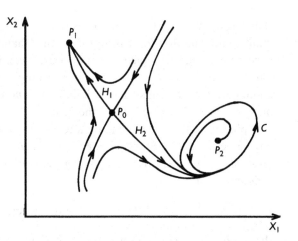

Fig. 3.4 Three types of invariant sets coexisting in a two-dimensional flow: fixed points (P_0, P_1, P_2), a closed curve C free of fixed points and a pair of heteroclinic trajectories H_1 and H_2 connecting invariant sets.

configurations of one-dimensional manifolds can arise such as curves joining fixed points (referred to as *heteroclinic trajectories*) or curves leaving a fixed point and subsequently returning to it (*homoclinic trajectories* or *separatrix loops*). A typical example is shown in Fig. 3.4. Notice that one disposes of the full classification of the phase space portraits of two-dimensional dynamical systems (Andronov *et al.*, 1966).

C Three variables and beyond

Since $n = 3$, one may have invariant manifolds with $d = 0, 1,$ or $2; d = 0$ corresponds again to fixed points and $d = 1$ to closed curves free of fixed points, or to curves joining fixed points familiar from the discussion in Sections 3.3A, and 3.3B. It is worth noticing, however, that contrary to the $n = 2$ case a closed curve (or a homoclinic trajectory) may now be of the many-circuit type without violating the uniqueness theorem.

Let us now look for some genuinely new possibilities in the form of two-dimensional invariant manifolds free of fixed points. Among the familiar surfaces in elementary three-dimensional geometry the cylinder, the cone, the hyperboloid or the paraboloid are to be excluded since they are not compact. The sphere (and all other surfaces homeomorphic to it) is an allowed possibility. However, as shown by Poincaré it may be mapped in a unique fashion onto a plane tangent to its 'south' pole by means of a projection using the corresponding 'north' pole as center. There is, therefore, no specifically new behavior expected in connection with this type of invariant manifold.

Less familiar, but very important for our purposes are the simple torus and the k-fold torus (torus with $k \geq 2$ holes). It is shown in topology

(Patterson, 1959) that a k-fold torus must possess at least $2k - 2$ fixed points, implying that there is no torus other than the simple one free of fixed points. Fig. 3.5 depicts a simple torus. It can be constructed by first joining together two opposite edges of a rectangle, a process giving rise to a finite cylinder. By subsequently joining the two remaining edges of the original rectangle one produces the torus.

A phase space point moving on a torus can be parameterized by two angular coordinates θ and ϕ (see Fig. 3.5). As a result the motion is *biperiodic*. If the periods T_1, T_2 along the two angular coordinates are rationally related,

$$T_1/T_2 = p/q \quad (p, q \text{ integers}) \tag{3.9}$$

the motion will reduce to a periodic one and will be represented by a closed curve winding p and q times on the torus, respectively, along the two angular directions. But if no relation of the kind of eq. (3.9) can be found the motion will be *quasi-periodic* and will be represented by a helix winding on the torus without ever closing to itself and without any self-intersection. The torus topology allows this flexibility, contrary to the plane or sphere topologies. We conclude that two-dimensional tori embedded in a three-dimensional space constitute the natural prototype of quasi-periodic behavior. As stressed in Chapter 1 such a behavior is encountered in large classes of nonlinear systems under nonequilibrium constraints.

The above arguments carry through in higher-dimensional dynamical systems, where 2, 3, ... up to $(n - 1)$-tori can be embedded in an n-dimensional space. The natural question to be raised is whether this family exhausts the list of allowable invariant manifolds. Although a full classification of compact surfaces in a multi-dimensional space is still an open problem, one can imagine *a priori* other possibilities of compact, invariant nonintersecting sets such as the Möbius band in three dimensions or the Klein bottle in four dimensions. Such possibilities are sometimes realized – for instance, in connection with homoclinic behavior in certain multivariate dynamical systems (Wiggins, 1990). However, the behavior of trajectories on such manifolds does not seem to introduce new elements beyond the ones considered above.

Fig. 3.5 Successive steps in the construction of a torus (after Patterson (1959)).

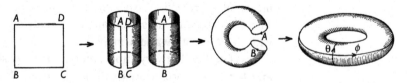

D Fractal manifolds

Our exploration of the geometry of phase space has led us to identify the prototypes of stationary behavior (fixed point), periodic behavior (closed curve) and quasi-periodic behavior (torus). We are still not in possession of the prototype of chaotic behavior which, as shown in Chapter 1, is abundant in large classes of natural systems. In this subsection we show that the flexibility afforded by a phase space of sufficiently large dimensionality, more precisely $n \geq 3$, allows one to envision manifolds of a new kind capable of carrying chaotic nonself-intersecting phase space trajectories.

In looking for invariant manifolds other than the manifolds of conventional geometry referred to in the previous subsections one is implicitly raising the question of the existence of objects that are neither points ($d = 0$), nor curves ($d = 1$), surfaces ($d = 2$) or hypersurfaces ($d = 3$ and beyond). Modern mathematics shows that such objects do indeed exist. They are 'intermediate' between two conventional manifolds of dimension d and $d + 1$ in the sense that although their Lebesgue measure (or more plainly their 'volume') in the $(d + 1)$-dimensional space is zero, they are nevertheless 'larger' in some well-defined sense (to be specified below) than sets constituting the d-dimensional manifold.

The canonical algorithm to construct such *fractal sets* (Mandelbrot, 1977; Feder, 1988; Schröder, 1991) mimics a process of successive fragmentation, a simple example of which is depicted in Fig. 3.6. From a d-dimensional object of characteristic size ℓ, N conformal copies of reduced size ℓr ($0 < r < 1$) are generated. From each of them N further reduced copies of size ℓr^2 are generated, and so forth. In the limit of infinite fragmentation one arrives at an infinite set of points (or, at most, of $(d - 1)$-dimensional objects) whose dimensionality should normally be zero (or, at most, $d - 1$). Yet one intuitively feels that this set has more content than that. To implement this idea in a quantitative fashion one introduces the concept of *fractal dimension*, D_0. For the simple example of Fig. 3.6 this can be done most conveniently by counting how the number

Fig. 3.6 Construction of a fractal by successive fragmentation of an object of size ℓ into N^n copies of reduced size ℓr^n ($0 < r < 1$), $n = 1, 2, \ldots$ (after Tel (1987)).

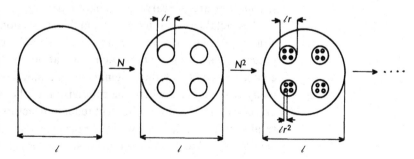

N_ε of members of the set of a given size ε varies with ε in the limit of small ε,

$$N_\varepsilon = \left(\frac{\varepsilon}{\ell}\right)^{-D_0}, \quad \varepsilon \to 0 \tag{3.10}$$

Notice that this reproduces the topological dimensions of the familiar manifolds of Euclidean geometry. Turning now to the example of Fig. 3.6, we see that there is $N_\ell = 1$ object of size ℓ, $N_{\ell r} = N$ objects of size ℓr, $N_{\ell r^2} = N^2$ objects of size ℓr^2 etc .... Substituting into (3.10) one finds

$$N^n = r^{-nD_0}$$

or, taking logarithms and simplifying by n,

$$D_0 = \frac{\ln N}{\ln (1/r)} \tag{3.11}$$

A familiar illustration of the above procedure is the celebrated Cantor set in which the unit interval is subdivided into three segments of length $1/3$ from which the middle one (without its boundaries) is deleted. Each of the two remaining segments is next divided into three equal parts with the open middle once again deleted, and so forth. Setting $N = 2$ and $r = 1/3$ in (3.11) one obtains $D_0 = \ln 2/\ln 3 \approx 0.63$. The Cantor set is therefore in this sense intermediate between a point ($d = 0$) and a line ($d = 1$). It is a *fractal object*, in the sense that its generalized dimension is strictly larger than its topological dimension $d = 0$.

Having established the existence of fractal sets one can now imagine invariant manifolds consisting of an infinity of sheets such that a section taken transversally to the sheets is a Cantor-like set visited successively by the phase space trajectory in a nonreproducible fashion as time follows its course. In a two-dimensional space this cannot be achieved without self-intersection of the trajectories. But in a space of dimensionality greater than or equal to three, such configurations become possible, typically by a process of successive *foldings* produced as the trajectory is winding on the manifold (Fig. 3.7). Such objects, which we shall refer to as *strange sets* are the carriers of chaotic behavior.

The fractal sets introduced so far in this subsection enjoy the property of strict *self-similarity*, since the presence of a single scale r in the reduction process guarantees that in each generation the resulting objects are conformal copies of the original one. A more typical case is that of many-scale fractals, where the subdivision of the original object into N objects is carried out with different reduction factors $r_i, i = 1, \ldots, N$. This results in a complex, highly inhomogeneous structure referred to as a *multifractal*. One can still define for such objects a fractal dimension D_0, but this dimension now appears as a sort of statistical average. Indeed, setting

$$N_\varepsilon = \sum_{i=1}^{N} N_i(\varepsilon)$$

applying eq. (3.10) to both sides and recognizing that $N_i(\varepsilon) = N(\varepsilon/r_i)$ we obtain

$$\left(\frac{\varepsilon}{\ell}\right)^{-D_0} = \sum_{i=1}^{N} \left(\frac{\varepsilon}{r_i\ell}\right)^{-D_0}$$

or finally

$$\sum_{i=1}^{N} r_i^{D_0} = 1 \qquad (3.12)$$

Fig. 3.7 The folding process leading to fractal sets in a phase space flow (after Abraham and Shaw (1985)).

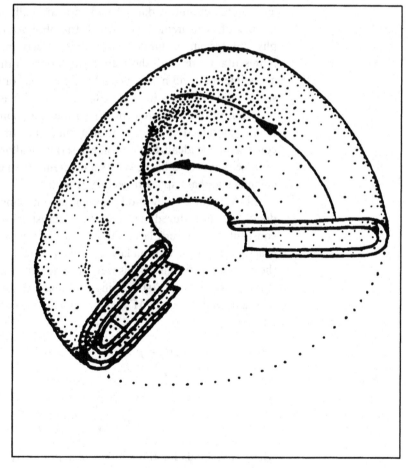

The new aspect is now that, in addition to D_0, one needs new parameters to characterize the inhomogeneity of the set. These so-called generalized dimensions, D_q (Feder, 1988) play in the theory a role analogous to the moments of a probability distribution. We do not develop this point further here but rather refer the reader to the abundant specialized literature.

3.4 Conservative and dissipative systems. Attractors

As we saw previously, the solution of the evolution equations of a dynamical system (eq. (3.6)) constitutes a well-posed problem in the sense that, under mild conditions, a complete specification of the state $\mathbf{X} = (X_1, \ldots, X_n)$ at any one time allows prediction of the state at all later times. In practice, such a complete specification amounts to disposing of an infinite amount of data. This is operationally meaningless since the process of measurement, by which the observer communicates with a physical system, is limited by a finite precision. In the phase space representation this will show up through the fact that the experimentally accessible state will not be given by a point but rather by a volume $\Delta\Gamma_0$ surrounding such a point, whose linear dimension is roughly given by the precision of the measurement. From the standpoint of the observer all points contained in $\Delta\Gamma_0$ represent the same (macroscopic) state; in contrast, in the idealized point-like description afforded by eqs. (3.6) each of these points is to be viewed as a different initial condition from which emanates a phase space trajectory (Fig. 3.8).

Gibbs invented a new mode of approach enabling one to cope with this duality. He introduced the concept of a statistical *ensemble* (Gibbs, 1902) constituted by a very large number of identical systems, all subject to exactly the same evolution laws and external constraints, but differing in their initial conditions. In this view, the relevant quantity to be considered is the probability $p_{\Delta\Gamma}(t)$ of being in a phase space cell $\Delta\Gamma$ at time t or, taking the limit of small $\Delta\Gamma$ and introducing the corresponding probability density ρ,

$$p_{\Delta\Gamma}(t) = \rho(X_1, \ldots, X_n, t)\, \mathrm{d}X_1 \ldots \mathrm{d}X_n \qquad (3.13a)$$

with

$$\rho = \frac{1}{N_{\text{tot}}} \lim_{\Delta\Gamma \to 0} \frac{\Delta N}{\Delta\Gamma} \qquad (3.13b)$$

where ΔN stands for the number of states in $\Delta\Gamma$ and N_{tot} for the total number of phase space states available. The relevance of Gibbs' invention is becoming increasingly obvious in the light of the discovery of the complex behaviors surveyed in Chapter 1. For such behaviors, and particularly for the chaotic regime, the phase space motion becomes very complex, and it is no longer meaningful to argue in terms of individual trajectories. The probabilistic description underlying the idea of Gibbs' ensemble provides one with a valuable alternative for describing the evolution of complex systems, which will be developed further in Chapter 7.

In order to predict the probability of occurrence of particular values of the state variables $\{X_i\}$, one must set up an equation of evolution for ρ. The uniqueness theorem of solutions of (3.6) formulated in Section 3.2 ensures that the number of phase space trajectories emanating from a certain set of initial data is conserved. This entails that ρ, which plays the role of the density of trajectories, behaves like the mass density of a fluid (eq. (2.13)) provided that the physical space coordinates $\{r_i\}$ in eq. (2.13) are replaced by the phase space ones $\{X_i\}$ and the velocities $\{v_i\}$ by the evolution laws $\{F_i\}$. One thus obtains

$$\frac{\partial \rho}{\partial t} + \sum_i \frac{\partial}{\partial X_i}(\rho F_i) = 0 \tag{3.14}$$

Performing the derivation of the product in (3.14) and introducing the hydrodynamic derivative by analogy to eq. (2.14b) one can further transform this equation into

Fig. 3.8 Phase space trajectories emanating from the representative phase space points inside an initial volume $\Delta\Gamma_0$.

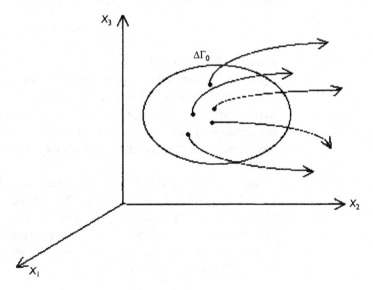

$$\frac{d\rho}{dt} = \frac{\partial \rho}{\partial t} + \sum_i F_i \frac{\partial \rho}{\partial X_i} = -\rho \ \mathrm{div} \ \mathbf{F}$$

or

$$\frac{d \ln \rho}{dt} = -\mathrm{div} \ \mathbf{F} \tag{3.15}$$

Integrating this relation formally from 0 to t one obtains

$$\frac{1}{t} \ln \frac{\rho_t}{\rho_0} = -\frac{1}{t} \int_0^t dt' \ \mathrm{div} \ \mathbf{F} \equiv -\overline{(\mathrm{div} \ \mathbf{F})}_t \tag{3.16a}$$

where the bar denotes the time average and $\rho_t \equiv \rho(.\,, t)$, $\rho_0 = \rho(.\,, 0)$. On the other hand, introducing expression (3.13b) for ρ and taking into account the fact that by the uniqueness theorem the number of states ΔN is conserved, one can further transform the above relation to

$$\frac{1}{t} \ln \frac{\Delta\Gamma_t}{\Delta\Gamma_0} = \overline{(\mathrm{div} \ \mathbf{F})}_t \tag{3.16b}$$

Eqs. (3.16) are the starting point of the classification of dynamical systems into the two important classes of conservative and dissipative systems.

A Conservative systems

By definition a dynamical system is called conservative if

$$\mathrm{div} \ \mathbf{F} = 0 \tag{3.17}$$

It follows from (3.15) that $d\rho/dt = 0$, which is nothing but the Liouville equation familiar from classical statistical mechanics (Prigogine, 1962). Alternatively, from eq. (3.16b) it follows that $\Delta\Gamma_t = \Delta\Gamma_0$, that is to say, the measure of a phase space volume is conserved during the evolution of a conservative system. This is the content of Liouville's theorem, another important result of classical dynamics (Goldstein, 1959; Lichtenberg and Lieberman, 1983).

An important class of conservative systems is the Hamiltonian systems. Indeed, using Hamilton's equations (2.2) one can see straightforwardly that eq. (3.17) is fulfilled identically. However, the concept of a conservative dynamical system is more general than that of a Hamiltonian system since, for one thing, the number of variables involved need not be even as in the Hamiltonian case.

B Dissipative systems

In a nonconservative system $\operatorname{div} \mathbf{F} \neq 0$. We will define the class of dissipative systems by the more restrictive condition

$$\overline{(\operatorname{div} \mathbf{F})}_t < 0, \quad t \geq t_0 \tag{3.18}$$

From eq. (3.16b) it follows that for such systems there is, on the average, contraction of phase space volume beyond a certain interval of time,

$$\Delta \Gamma_t < \Delta \Gamma_0, \quad t \geq t_0 \tag{3.19}$$

This entails that in the limit $t \to \infty$ the trajectories of a dissipative system initially emanating from a certain phase space volume $\Delta \Gamma_0$ will tend to a subset of phase space of zero volume, i.e. a subset whose dimension will be strictly less than the phase space dimension. This set is referred to as an *attractor* of the dynamical system. Since by the above definition attractors are invariant manifolds, one can apply to them the analysis of Section 3.3. On this basis one expects to find in dissipative systems zero-dimensional (fixed point) attractors; one-dimensional (periodic) attractors, referred to as *limit cycles*; two-dimensional and higher-dimensional (quasi-periodic) attractors in the form of invariant tori; and fractal attractors. The largest, compact, invariant set having the attracting property is referred to as the *universal attractor*. Clearly, every evolution of a dissipative system from a certain time on remains on this universal attractor for ever.

From the above short discussion one can realize why physically relevant nonconservative systems must obey the more restrictive condition (3.18). Indeed, had the inequality sign been reversed, the system would eventually escape to infinity. Such an explosive behavior is ruled out in physical systems, which are characterized by finite values of energy, mass and other macroscopic observables. Notice that the phase space volume contraction need not hold everywhere: it suffices to have this condition satisfied on average. This is what happens in the Brusselator model (eq. (2.44)), where $F_X = k_1 A - (k_2 B + k_4)X + k_3 X^2 Y$, $F_Y = k_2 B X - k_3 X^2 Y$ and $\operatorname{div} \mathbf{F} = -(k_2 B + k_4) + k_3 X (2Y - X)$. In contrast, certain systems like the Lorenz model (eqs. (3.4)) satisfy the more strict dissipativity condition of $\operatorname{div} \mathbf{F}$ being negative everywhere: $\operatorname{div} \mathbf{F} = -(\sigma + b + 1)$. A similar property holds when a linear-damping term of the form $-\gamma d\theta/dt$ is added to the equation of evolution of a conservative oscillator like the Duffing oscillator, eq. (2.9a).

3.5 Stability

Our next objective will be to characterize the invariant manifolds introduced in the preceding sections in a more detailed manner.

Consider a system – dissipative or conservative – evolving according to eqs. (3.6). We suppose that by a mechanism that we need not specify here, the system has reached after a certain lapse of time a 'reference' state $\mathbf{X}_s$ on an invariant manifold. In principle, by the very definition of invariant manifold, one expects that the system will remain therein for ever and will undergo a dynamical behavior dictated by the particular type of manifold considered.

In actuality, a real-world system never stays in a single state as time varies. To begin with, most systems are in contact with a complex environment with which they exchange matter, momentum and energy in a practically unpredictable manner. In addition, most of the systems encountered in nature intrinsically generate their own variability in the form of *thermodynamic fluctuations*. As a result the instantaneous state $\mathbf{X}(t)$ will continuously deviate from $\mathbf{X}_s$ by an amount $\mathbf{x}(t)$, referred to as the *perturbation*,

$$\mathbf{X}(t) = \mathbf{X}_s + \mathbf{x}(t) \tag{3.20}$$

We are interested in the response $\mathbf{x}(t)$ of a dynamical system to an initial deviation $\mathbf{x}(0)$ of the kind defined above. In particular we want to know whether, upon the action of the perturbation, the system will remain close to the reference state or on the contrary will deviate significantly from it. It is here that stability, one of the key concepts of the theory of dynamical systems, enables one to envisage things in a clearcut manner.

Let us formulate the problem in phase space. We denote (Fig. 3.9) by U_ε

Fig. 3.9 Geometric
view of stability.

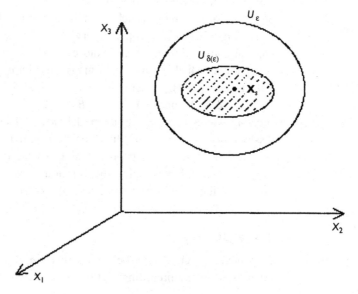

and $U_{\delta(\varepsilon)}$ two regions surrounding $\mathbf{X}_s$ (represented as a point in Fig. 3.9) whose characteristic sizes are given respectively by ε and δ (with δ generally depending on ε). We adopt the following definitions:

$\mathbf{X}_s$ is *stable in the sense of Lyapunov* if, for any given neighborhood U_ε of $\mathbf{X}_s$ there exists a certain neighborhood $U_{\delta(\varepsilon)}$ such that any trajectory emanating from the interior of $U_{\delta(\varepsilon)}$ never leaves U_ε.
$\mathbf{X}_s$ is *unstable* if no such neighborhood $U_{\delta(\varepsilon)}$ can be found.
$\mathbf{X}_s$ is *asymptotically stable* if it is stable and if, in addition, any trajectory emanating from the interior of $U_{\delta(\varepsilon)}$ tends to $\mathbf{X}_s$ as $t \to \infty$.

To this geometric view of stability one can associate an analytic formulation, provided that one endows phase space with a *norm*, $|.|$. Specifically one has

Lyapunov stability if, for any $\varepsilon > 0$, there exists a $\delta(\varepsilon) > 0$ such that for any $\mathbf{X}(0)$ with $|\mathbf{X}(0) - \mathbf{X}_s| < \delta$, one has $|\mathbf{X}(t) - \mathbf{X}_s| < \varepsilon$ for all $t \geq 0$. Asymptotic stability if, in addition, $|\mathbf{X}(t) - \mathbf{X}_s| \to 0$ as $t \to \infty$.

In this latter case $\mathbf{X}_s$ will be an attractor of the dynamical system. Clearly then, asymptotic stability can hold only in dissipative systems.

The great value of the above definitions is to reflect the intuitive idea that stability, and especially asymptotic stability, is a property that makes, in a way, a given state physically legitimate. Indeed, in view of the ubiquity of perturbations and fluctuations in nature, a state lacking stability would become unobservable after a certain lapse of time.

A second most important point is that stability allows one to formulate systematically, and in quantitative terms, the onset of complex behavior. Specifically, let $\mathbf{X}_s$ first represent one of the fixed points of the dynamical system (eqs. (3.6)),

$$\mathbf{F}(\mathbf{X}_s, \lambda) = 0 \qquad (3.21)$$

The determination of these particular invariant sets reduces to a problem of algebra that can, in principle, be solved. If $\mathbf{X}_s$ happens to be stable the system's behavior will essentially be determined by the knowledge of $\mathbf{X}_s$, and the classification problem raised in the first sections of this chapter will be solved. Conversely, the onset of behaviors of a type more complex than the fixed point one will be signaled by the failure of the stability of the fixed points, whereupon the system will be bound to evolve toward invariant sets of a new type.

In view of the above it now becomes crucial to devise methods for *testing* the stability of a given invariant set, say a fixed point. In some exceptional cases this can be done as a straightforward application of the very definition of stability. As an example consider the classical harmonic oscillator to which we have already alluded as a limiting case of the hoop problem (eq. (2.9a), $\mu < 0$, cubic term neglected). Introducing the velocity $v = d\theta/dt$ and the angular frequency $\omega_0^2 = -\mu$ one can write the equations of motion as

$$\left.\begin{array}{l} d\theta/dt = v \\ dv/dt = -\omega_0^2\theta \end{array}\right\} \qquad (3.22)$$

which define in the phase space a continuum of ellipses surrounding the equilibrium point $\theta = v = 0$ (Fig. 3.10). We want to test the stability of this fixed point. To this end we consider a neighborhood U_ε in the form of a square of side ε. Among the integral curves of the system there exists one that is tangent to the vertical sides of the square and contained entirely in its interior. We choose the part of the phase plane inside this curve as our neighborhood $U_{\delta(\varepsilon)}$. Consider now an initial condition inside this neighborhood. By virtue of the uniqueness theorem a trajectory emanating from this initial condition will never cross another integral curve (not containing a fixed point) and will thus remain, by construction, inside the neighborhood U_ε. We conclude that the fixed point (0,0) is stable in the sense of Lyapunov. Such points, surrounded by a continuum of elliptic trajectories, are also referred to as *elliptic points*, the motions around them being qualified as *stable motions*.

A very different case corresponds to having a positive coefficient μ in the

Fig. 3.10 Proof of Lyapunov stability of the fixed point of the harmonic oscillator, eqs. (3.22).

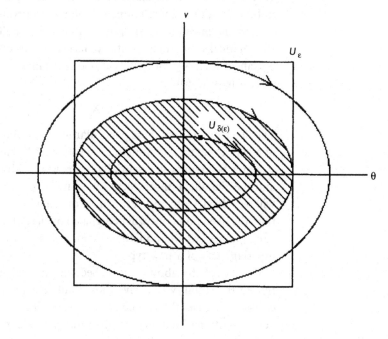

hoop problem (eq. (2.9a)). Neglecting for a moment again the cubic term one obtains

$$\left.\begin{array}{l} d\theta/dt = v \\ dv/dt = \omega_0^2\theta \end{array}\right\} \tag{3.23}$$

which now define in the phase space ($x = v + \omega_0\theta$, $y = v - \omega_0\theta$) a continuum of hyperbolas around the equilibrium point (0,0). The motion consists (Fig. 3.11) of a contraction along the y direction and an expansion along the x direction. Arguing as above one will find that the equilibrium point is unstable (Section 1.2, Fig. 1.2, branch (a')). We call such points *hyperbolic points*, the motions around them being qualified as *unstable motions*.

In both eqs. (3.22) and (3.23) the phase space volume is conserved, since condition (3.17) is satisfied. But while in the first case its shape remains essentially unchanged, in the second case it is highly deformed since an initial volume in the form of a square will eventually become a rectangle whose horizontal side will tend to infinity and whose vertical one to zero. We already find in this simple example the ingredients of more complex phenomena related to chaos that will be studied in more detail later.

So far we have not used explicitly the property that eqs. (3.6) are autonomous. We shall now inquire into the repercussions of this property in stability. Let $X_i(t)$ be a solution of eqs. (3.6). It is then clear that any function $X_i(t + \tau)$ where τ is an arbitrary constant (the phase) is still a solution of the same equations. In other words, autonomous systems exhibit the property of translational invariance in time. These infinitely

Fig. 3.11 Deformation of a phase space volume in a dynamical system possessing a hyperbolic point (eqs. (3.23)).

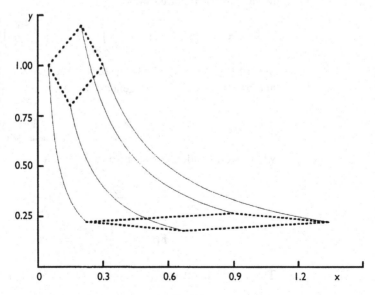

many solutions, differing from each other by the phase, define an orbit, C in phase space.

We say that C is orbitally stable if, given $\varepsilon > 0$, there exists a $\delta > 0$ such that if $\mathbf{X}_0$ is a representative point of another orbit within a distance δ from C at time 0, then its image $\mathbf{X}$ remains within a distance ε from C for any $t > 0$. Otherwise, C is orbitally unstable. If C is orbitally stable and the distance between $\mathbf{X}$ and C tends to zero as $t \to \infty$, C is asymptotically orbitally stable.

Lyapunov stability and orbital stability should not be confused, see problems 3.6 and 3.7.

3.6 The principle of linearized stability

The straightforward assessment of stability on the basis of the definitions laid down in Section 3.5 works only in rather exceptional cases. A systematic, analytically tractable algorithm is clearly needed to tackle more general and more complex cases.

The starting point is to substitute eq. (3.20) into (3.6). Utilizing the fact that the reference state $\mathbf{X}_s$ is itself a particular solution of these latter equations one obtains

$$\mathrm{d}\mathbf{x}/\mathrm{d}t = \mathbf{F}(\mathbf{X}_s + \mathbf{x}, \lambda) - \mathbf{F}(\mathbf{X}_s, \lambda) \tag{3.24}$$

We assume that $\mathbf{F}$ can be Taylor expanded in formal power series of $\mathbf{x}$ around $\mathbf{X}_s$. This is always possible if $\mathbf{F}$ has a polynomial structure and leads then to a finite number of terms, but may require further conditions in more intricate situations:

$$\mathbf{F}(\mathbf{X}_s + \mathbf{x}, \lambda) = \mathbf{F}(\mathbf{X}_s, \lambda) + \left(\frac{\delta \mathbf{F}}{\delta \mathbf{X}}\right)_{\mathbf{X}_s} \cdot \mathbf{x} + \frac{1}{2}\left(\frac{\delta^2 \mathbf{F}}{\delta \mathbf{X} \delta \mathbf{X}}\right)_{\mathbf{X}_s} \cdot \mathbf{xx} + \ldots \tag{3.25}$$

Substituting into eq. (3.24) one sees that the inhomogeneous term cancels and one is left with a homogeneous problem for the perturbation vector $\mathbf{x}$:

$$\mathrm{d}\mathbf{x}/\mathrm{d}t = \underset{\text{linearized part}}{\mathscr{L}(\lambda) \cdot \mathbf{x}} + \underset{\text{nonlinear contributions}}{\mathbf{h}(\mathbf{x}, \lambda)} \tag{3.26}$$

where we introduced the short hand notation

$$\mathscr{L}(\lambda) = \left(\frac{\delta \mathbf{F}}{\delta \mathbf{X}}\right)_{\mathbf{X}_s} \tag{3.27a}$$

$$\mathbf{h}(\mathbf{x}, \lambda) = \frac{1}{2}\left(\frac{\delta^2 \mathbf{F}}{\delta \mathbf{X} \delta \mathbf{X}}\right)_{\mathbf{X}_s} \cdot \mathbf{xx} + \ldots \tag{3.27b}$$

The linear operator $\mathscr{L}(\lambda)$ is simply the Jacobian matrix of $\mathbf{F}$ evaluated at the reference state, whereas $\mathbf{h}(\mathbf{x}, \lambda)$ contains contributions that are nonlinear in $\mathbf{x}$ and has, therefore, the property $\mathbf{h}(\mathbf{x}, \lambda) = O(|\mathbf{x}|^2)$, or if the series (3.27b) can be differentiated term by term, $(\partial \mathbf{h}/\partial \mathbf{x}) \to 0$ as $|\mathbf{x}| \to 0$. If the reference state $\mathbf{X}_s$ is time-independent (fixed point) then both $\mathscr{L}$ and $\mathbf{h}$ will be free of any explicit time dependence.

Comparing (3.26) and (3.6) we see that the former is an equivalent version of the latter in which the origin of coordinates in phase space has been placed on $\mathbf{X}_s$. The (trivial) solution $\mathbf{x} = 0$ of the (homogeneous) system of eq. (3.26) is, clearly, the analog of the reference state $\mathbf{X}_s$ of the initial problem defined by eqs. (3.6).

In dynamical systems involving a finite number of degrees of freedom $\mathbf{h}(\mathbf{x}, \lambda)$ is a vector in phase space, whereas $\mathscr{L}(\lambda)$ is a $n \times n$ matrix whose elements are given by $\mathscr{L}_{ij} = (\partial F_i/\partial X_j)_{\{X_{js}\}}$ $(i, j = 1, \ldots, n)$. For instance, in the Brusselator model in a well-stirred medium (eqs. (2.44)), taking for simplicity $k_i = 1$ one has

$$F_X = A - (B + 1)X + X^2 Y, \quad F_Y = BX - X^2 Y, \qquad (3.28a)$$

and a single fixed point $\mathbf{X}_s = (A, B/A)$. Choosing this point as a reference state one finds straightforwardly that

$$\mathscr{L} = \begin{pmatrix} B - 1 & A^2 \\ -B & -A^2 \end{pmatrix}, \quad \mathbf{h} = \begin{pmatrix} \dfrac{B}{A}x^2 + 2Axy + x^2 y \\ -\dfrac{B}{A}x^2 - 2Axy - x^2 y \end{pmatrix}$$

$$(3.28b)$$

Eq. (3.26) constitutes still a highly nonlinear problem which, as a rule, is as intractable as the original problem, eqs. (3.6). At this point, however, a most important result can be invoked to enable further progress. This theorem, also known as the *principle of linearized stability*, compares the stability properties of the following two problems:

The original, fully nonlinear problem (eqs. (3.26)).
The 'auxiliary' linearized problem, in which higher order terms are omitted,

$$d\mathbf{x}/dt = \mathscr{L}(\lambda) \cdot \mathbf{x} \qquad (3.29)$$

It stipulates the following:

If the trivial solution $\mathbf{x} = 0$ of the linearized problem (eq. (3.29)) is asymptotically stable, then $\mathbf{x} = 0$ (or equivalently $\mathbf{X} = \mathbf{X}_s$) is an asymptotically stable solution of the nonlinear problem, eqs. (3.26) or (3.6). If the trivial solution $\mathbf{x} = 0$ of the linearized problem is unstable, then

$x = 0$ (or equivalently $\mathbf{X} = \mathbf{X}_s$) is an unstable solution of the nonlinear problem.

The theorem is unable to provide information in the case in which the trivial solution of (3.29) is Lyapunov stable but not asymptotically stable. Still, it is of the utmost value since it reduces the passage from stability to instability, one of the fundamental problems of dynamical systems theory, to a linear problem – a much more traditional and tractable problem of analysis.

Intuitively the theorem seems reasonable since, after all, stability reflects the response of a system to small perturbations for which the expansion of eq. (3.25) can be truncated to its first significant (here linear) term. Rigorous demonstrations can be found in the abundant mathematical literature. To give a flavor of the argument we reproduce in Appendix A1 the proof for a one-variable system and illustrate it on the logistic equation (eq. (2.51)). In the multivariate case two versions are usually encountered. The most traditional one (Nemytskii and Stepanov, 1960; Cesari, 1963; Sattinger, 1972) amounts to the statement:

If all eigenvalues of $\mathscr{L}$ in (3.26) have negative real parts then $\mathbf{x} = 0$ is an asymptotically stable solution. If some eigenvalues of $\mathscr{L}$ have positive real parts, then $\mathbf{x} = 0$ is unstable.

In this form the theorem is proved either by the method of successive approximations or by the use of Lyapunov functions (the so-called Lyapunov's second method).

A more far-reaching formulation (Arnol'd, 1980; Guckenheimer and Holmes, 1983; Arrowsmith and Place, 1990), which actually goes back to Poincaré and is known in its modern version as the Hartman–Grobman theorem, is as follows:

If $\mathscr{L}(\lambda)$ has no zero or purely imaginary eigenvalues then there is a homeomorphism defined in some neighborhood of $\mathbf{X}_s$ in R^n locally taking orbits of the nonlinear flow of (3.26) to those of the linear flow of (3.29). The homeomorphism preserves the sense of orbits and can also be chosen to preserve parameterization by time.

The proof of this important result is surprisingly simple and can be found in Arnol'd (1980). When $\mathbf{X}_s$ is such that $\mathscr{L}(\lambda)$ has the above property, $\mathbf{X}_s$ is called a *hyperbolic or nondegenerate* fixed point. In this case, then, the linear and nonlinear flows are topologically equivalent. A simple illustration on the Brusselator model, eq. (2.44), is depicted in Fig. 3.12.

Fig. 3.12 Illustration of the Hartman–Grobman theorem on the Brusselator model, eq. (2.44), for $k_i = 1$, $A = 2$. In (a) $B = 4.5$ and the fixed point is a stable focus. The trajectories of the linearized system (dashes) and of the nonlinear system (full lines) spiral toward the attractor. In (b) $B = 5$ and the system is in a state of marginal stability. The trajectories of the linearized system are closed curves surrounding the fixed point and are topologically different from those of the non-linear system, which spiral toward the fixed point.

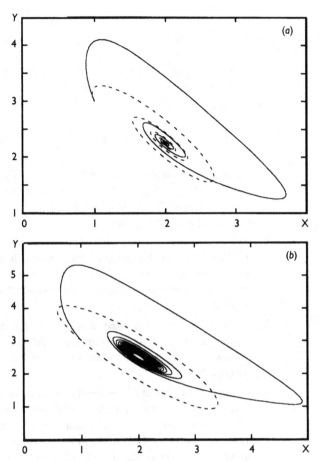

Problems

3.1 Let $Y(X)$ be the phase space trajectory of the dynamical system $dX/dt = f(X, Y)$, $dY/dt = g(X, Y)$. Derive an expression for the time needed to evolve from (X_0, Y_0) to (X_1, Y_1) in terms of f or g and the equation of the trajectory.

3.2 The definitions of stability of Section 3.5 apply as well to any reference solution $\mathbf{X}_s(t)$ of the evolution equations, not necessarily lying on an invariant manifold. Prove that every solution of $dx/dt = -\lambda x$, $\lambda > 0$ is asymptotically stable for $t > 0$. What happens for $t < 0$?

3.3 Write out the explicit form of eq. (3.15) for the one-variable system $dx/dt = -\lambda x$, $\lambda > 0$. Solve the initial value problem and the corresponding eigenvalue problem. Comment on the results in connection with the stability of the state $x = 0$.

3.4 Similarly for the harmonic oscillator model, eq. (3.22).

3.5 Stability versus boundedness (Cesari, 1963). (*a*) Every solution of
$dx/dt = 1$ is unbounded. Show that any one of these solutions, chosen as
a reference state, is nevertheless stable in the sense of Lyapunov. (*b*) The
solutions of the equation $dx/dt = x - x^3$ are bounded. Show that $x = 0$
is unstable and $x = \pm 1$ are asymptotically stable.

3.6 The system

$$\frac{dx}{dt} = -y(x^2 + y^2), \quad \frac{dy}{dt} = x(x^2 + y^2)$$

has the two-parameter family of solutions $x = C_1 \cos(C_1^2 t + C_2)$,
$y = C_1 \sin(C_1^2 t + C_2)$. Show that all solutions but $x = y = 0$ are
Lyapunov unstable, but that all integral curves (compact invariant
manifolds) are orbitally stable.

3.7 Noninvariance of stability with respect to a coordinate change (Cesari,
1963). Find the evolution equations generated from those of Problem 3.6
by performing the change of variables $x = r \cos(r^2 t + \psi)$, $y = r \sin$
$(r^2 t + \psi)$. Show that all solutions of the system for the new variables
(r, ψ) are Lyapunov stable.

3.8 Using the results summarized in Section 2.1 prove that any integrable
system can be transformed by an appropriate change of variables to a
system all of whose solutions are Lyapunov stable. What are these new
variables?

3.9 Consider a dynamical system on the two-d torus defined by the
equations $d\theta/dt = F(\theta\ \phi)$, $d\phi/dt = \alpha F(\theta, \phi)$ where α is a positive
irrational number, and $F(\theta, \phi)$ a continuous function satisfying the
Lipschitz condition, 2π-periodic in the arguments θ and ϕ, and positive
everywhere except at $(0,0)$ where $F(0,0) = 0$. Describe the motion
generated by these equations with special emphasis on stability and
compare it with the uniform motion generated by $d\theta/dt = 1$, $d\phi/dt = \alpha$.

3.10 Show that the system of coupled logistic equations

$$\frac{dX_1}{dt} = kX_1 \left(1 - \frac{X_1 + X_2}{N} \right)$$

$$\frac{dX_2}{dt} = kX_2 \left(1 - \frac{X_1 + X_2}{N} \right)$$

admits a continuum of degenerate steady-state solutions.
Determine the stability of these solutions and draw the phase portrait.
Sketch an intuitive interpretation of the results from the standpoint of
population dynamics and evolution theory (Gause, 1934; Allen, 1975).

CHAPTER FOUR

Linear stability analysis of fixed points

4.1 General formulation

The objective of this chapter is to set up *quantitative* criteria of stability of the fixed points of a dynamical system. This will be possible thanks to the principle of linearized stability which, as we saw in Section 3.6, reduces stability to a linear problem, eqs. (3.29).

In a system subjected to time-independent constraints the evolution laws **F** (eqs. (3.6)) do not depend explicitly on time. Furthermore, if one is interested in the stability of fixed points the reference state $\mathbf{X}_s$ is time-independent and so is also, by virtue of eq. (3.27a), the linearized operator $\mathscr{L}(\lambda)$. It follows that eqs. (3.29) admit solutions that depend on time exponentially,

$$\mathbf{x} = \mathbf{u}e^{\omega t} \tag{4.1}$$

Substituting into eqs. (3.29) one finds that **u** and the *characteristic exponent* ω must satisfy the relations

$$\mathscr{L}(\lambda) \cdot \mathbf{u} = \omega \mathbf{u} \tag{4.2a}$$

or, in more explicit form,

$$\sum_j \mathscr{L}_{ij}(\lambda) u_j = \omega u_i \tag{4.2b}$$

In other words **u** and ω are, respectively, eigenvectors and eigenvalues of $\mathscr{L}(\lambda)$ and stability is thus reduced to an eigenvalue problem. An important point is that independently of the properties of **u**, which takes into account the structure of **x** as a vector in phase space, knowledge of the eigenvalue ω provides one with a full solution of the problem of stability. Indeed, separating ω into real and imaginary parts we have from (4.1)

$$|\mathbf{x}| \approx e^{(\mathrm{Re}\,\omega)t}\, e^{i(\mathrm{Im}\,\omega)t} \tag{4.3}$$

It follows that

if $\mathrm{Re}\,\omega < 0$, $|\mathbf{x}|$ is exponentially decreasing and hence the reference state $\mathbf{x} = 0$ (or $\mathbf{X} = \mathbf{X}_s$) is *asymptotically stable*;
if $\mathrm{Re}\,\omega > 0$ the perturbations grow exponentially and hence the reference state is *unstable*.

These two regimes, for which the principle of linearized stability applies, are separated by the regime where $\mathrm{Re}\,\omega = 0$. We call this borderline case between asymptotic stability and instability *marginal stability*. Notice that the occurrence of instability and marginal stability is compatible with both conservative and dissipative systems. In contrast asymptotic stability implies by necessity a contraction of phase space volumes and can therefore occur only in dissipative systems.

The eigenvalue problem, eqs. (4.2), allows us to understand better the paramount importance of the control parameter(s) λ. Indeed, a variation of λ induces a variation of $\mathscr{L}$ and, through it, of the eigenvalue ω. Two typical possibilities are depicted by curves (*a*) and (*b*) of Fig. 4.1.

In (*a*), ω crosses the λ-axis with a positive slope. This will be reflected by the fact that as λ increases, the system will switch from asymptotic stability to instability. As the reference fixed point will no longer be a physically legitimate solution for $\lambda > \lambda_c$, a qualitative change of behavior is to be expected when λ_c is crossed. For this reason we shall refer to λ_c as the *critical value* of the control parameter. In contrast, in (*b*) the real part of the eigenvalue remains negative for all values of λ: the fixed point is always

Fig. 4.1 Two typical dependences of the real part of the eigenvalue of the linearized operator, eqs. (4.2), versus the control parameter λ: (*a*) the reference state is asymptotically stable for $\lambda < \lambda_c$ and unstable for $\lambda > \lambda_c$; (*b*) the reference state remains asymptotically stable for all values of λ.

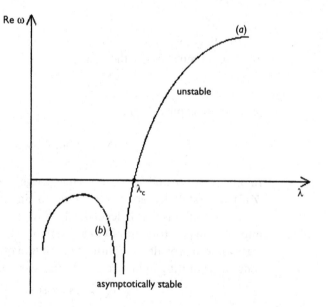

asymptotically stable and no qualitatively new regime is expected to arise spontaneously from the action of perturbations. In general a multivariable system possesses a whole spectrum of eigenvalues, some of which may behave as in (*a*) and others as in (*b*). For the transition to instability to take place it suffices that one eigenvalue behaves as in (*a*). The important point is that a given, well-defined system can switch between stability and instability according to the conditions to which it is subjected: stability is thus reduced to the parameter dependence of the solutions of the eigenvalue problem of the linearized operator.

It is clear by now that the central problem of stability theory is the determination of ω. An explicit calculation can be carried out from eq. (4.2b) which we write in the more suggestive form

$$\sum_{j=1}^{n} (\mathscr{L}_{ij}(\lambda) - \omega \delta_{ij}^{kr}) u_j = 0 \quad i = 1, \ldots, n \tag{4.4}$$

This set of homogeneous algebraic equations for $\{u_j\}$ admits a nontrivial solution provided that the determinant of the matrix of coefficients of $\{u_j\}$ vanishes. This gives rise to the *characteristic equation*

$$\det | \mathscr{L}_{ij}(\lambda) - \omega_m(\lambda)\delta_{ij}^{kr} | = 0 \tag{4.5}$$

where we have introduced the index m to account for the fact that eq. (4.5), which is an algebraic equation for ω_m, will in general admit several solutions. For typical values of the parameter one can legitimately expect that, unless $\mathscr{L}_{ij}$ has some remarkable symmetries built in, the eigenvalues ω_m will be distinct. The general solution of the linear system of eqs. (3.29) will then be given by

$$\mathbf{x} = \sum_{m=1}^{n} C_m \mathbf{u}_m e^{\omega_m t} \tag{4.6}$$

where C_m are integration constants determined by the initial conditions. Under the same conditions one knows from linear algebra that $\mathscr{L}$ can be diagonalized by a similarity transformation involving a nonsingular matrix $\mathbf{T}$,

$$\mathbf{D} = \mathbf{T}^{-1} \cdot \mathscr{L} \cdot \mathbf{T} \tag{4.7a}$$

where

$$D_{ij} = \omega_i \delta_{ij}^{kr} \quad i, j = 1, \ldots, n \tag{4.7b}$$

and the columns of $\mathbf{T}$ are given by the eigenvectors of $\mathscr{L}$. Operating on both sides of (3.29) by $\mathbf{T}^{-1}$ one can transform this system to

$$\mathbf{T}^{-1} \cdot d\mathbf{x}/dt = \mathbf{T}^{-1} \mathscr{L} \mathbf{T} \mathbf{T}^{-1} \cdot \mathbf{x}$$

or, introducing the new variables

$$\mathbf{z} = \mathbf{T}^{-1} \cdot \mathbf{x} \tag{4.8}$$

and taking (4.7) into account,

$$d\mathbf{z}/dt = \mathbf{D}\mathbf{z}$$

or finally

$$dz_i/dt = \omega_i z_i \quad i = 1, \ldots, n \tag{4.9}$$

The above results (eqs. (4.6)–(4.9)) can be generalized to include cases in which the matrix $\mathscr{L}$ has multiple eigenvalues. This happens frequently in the presence of symmetries or, more exceptionally, when the control parameters take some particular values. Actually one should distinguish between two types of multiplicities of an eigenvalue ω_m $(m = 1, \ldots, r)$ of a matrix $\mathscr{L}$:

the algebraic multiplicity μ_m, which is the multiplicity of ω_m as a root of the characteristic equation (4.5)
the geometric multiplicity ν_m (sometimes referred to as nullity), defined as the number of linearly independent vectors $\mathbf{g}$ such that

$$(\mathscr{L} - \omega_m \mathbf{1}) \cdot \mathbf{g} = 0 \tag{4.10}$$

Notice that one has necessarily $\nu_m \leq \mu_m$.

Given now an $n \times n$ matrix $\mathscr{L}$ of eigenvalues ω_m of algebraic and geometric multiplicities μ_m and ν_m respectively, one may show (Gantmacher, 1959) that there exist nonsingular matrices $\mathbf{T}$ such that

$$\mathbf{J} = \mathbf{T}^{-1} \cdot \mathscr{L}(\lambda) \cdot \mathbf{T} \tag{4.11a}$$

is the direct sum of N irreducible matrices $\mathbf{J}_s$ of orders n_s $(n_1 + \ldots + n_s + \ldots + n_N = n)$, referred to as *Jordan blocks*. The dimensionality n_s of $\mathbf{J}_s$ depends on the relation between the μ_m and ν_m. If $\mu_m = \nu_m$ then $n_s = 1$ and $\mathbf{J}_s = \omega_m$. If $\nu_m < \mu_m$ then $n_s > 1$. In this case each Jordan block can still be uniquely associated with an eigenvalue ω_m and has the following typical structure

$$\mathbf{J}_s = \begin{pmatrix} \omega_m & 1 & & & 0 \\ & \cdot & & & \\ & & \cdot & & \\ & & & \cdot & 1 \\ 0 & & & & \omega_m \end{pmatrix} \tag{4.11b}$$

if ω_m is real, or

$$J_s = \begin{pmatrix} \mathbf{R}_m & \mathbf{1} & & & 0 \\ & \cdot & & & \\ & & \cdot & & \\ & & & \cdot & \mathbf{1} \\ 0 & & & & \mathbf{R}_m \end{pmatrix}$$

(4.11c)

with

$$\mathbf{R}_m = \begin{pmatrix} \operatorname{Re} \omega_m & -\operatorname{Im} \omega_m \\ \operatorname{Im} \omega_m & \operatorname{Re} \omega_m \end{pmatrix}, \quad \mathbf{1} = \begin{pmatrix} 1 & 0 \\ 0 & 1 \end{pmatrix}$$

if ω_m is complex.

The relevance of these more sophisticated transformations will become obvious in Chapter 5, where explicit examples of Jordan blocks will be given. For now we come back to the generic case of distinct eigenvalues and survey, in the next three sections, a number of representative cases.

4.2 Systems involving one variable

Eqs. (3.6) reduce in this case to the single equation

$$\mathrm{d}X/\mathrm{d}t = F(X, \lambda)$$

(4.12a)

The phase space is one-dimensional and the linearized system, eqs. (3.29), becomes

$$\mathrm{d}x/\mathrm{d}t = (\partial F/\partial X)_{X_s} x$$

(4.12b)

The linearized operator $\mathcal{L}(\lambda)$ reduces to a number, also identical to the characteristic exponent ω,

$$\omega(\lambda) = \mathcal{L}(\lambda) = (\partial F/\partial X)_{X_s}$$

(4.13)

This number is by necessity real. Two typical situations can be realized, depending on whether $\omega(\lambda)$ is in the negative or in the positive half axis:

(a) ω is negative

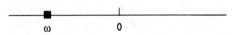

The phase portrait around the fixed point is given by

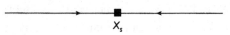

where the arrows account for the tendency of trajectories initially in a neighborhood of X_s to converge to it in the course of time. Since ω is real the convergence is monotonic.

(b) ω *is positive*

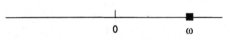

The phase space portrait is given by

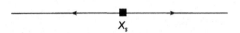

The trajectories now diverge from X_s. Since sheer explosion to infinity is not expected in a physical system, they will presumably tend to a new attractor. In a one-dimensional phase space the latter can only be another fixed point, located at some (typically finite) distance from X_s. The two cases *(a)* and *(b)* are separated by the critical case of marginal stability

$$\omega = \mathscr{L}(\lambda_c) = 0$$

for which linearization becomes inadequate.

A one-variable nonexplosive system is necessarily dissipative. Interesting examples, given in Chapter 2, are the Schlögl models (eqs. (2.45)), the Semenov model (eq. (2.47)) and the Verhulst model (eq. (2.51)). Taking the last as an illustration one has

$$F = kX(1 - X/N)$$

There are two fixed points $X_{s1} = 0$, $X_{s2} = N$, with

$$(\partial F/\partial X)_{X_{s1}} = k$$

$$(\partial F/\partial X)_{X_{s2}} = -k \qquad\qquad (4.14)$$

Recalling (eq. (2.50)) that k is an excess parameter expressing the difference between reproduction and death rates we see that the state of extinction is the unique physically acceptable state for $k < 0$ and is asymptotically stable in this range (see also Appendix A1). When $k > 0$ this state still exists, but becomes unstable. A new, nontrivial state $X_{s2} = N$ is born which 'inherits' the stability of X_{s1}. We see that one-variable systems can produce nontrivial behavior reminiscent of some of the experimental results surveyed in Chapter 1. The possibilities are, however, far more limited than for two-variable systems to which we turn next.

4.3 Systems involving two variables

The phase space is now two-dimensional and the linearized equations (3.29) take the form

$$\left.\begin{aligned} dx_1/dt &= \mathscr{L}_{11}x_1 + \mathscr{L}_{12}x_2 \\ dx_2/dt &= \mathscr{L}_{21}x_1 + \mathscr{L}_{22}x_2 \end{aligned}\right\} \quad (4.15)$$

Inserting (4.1) one may write the eigenvalue problem, eqs. (4.4), as

$$\left.\begin{aligned} (\mathscr{L}_{11} - \omega_m)u_{m1} + \mathscr{L}_{12}u_{m2} &= 0 \\ \mathscr{L}_{21}u_{m1} + (\mathscr{L}_{22} - \omega_m)u_{m2} &= 0 \qquad m = 1, 2 \end{aligned}\right\} \quad (4.16)$$

Here u_{m1}, u_{m2} are the two components of the eigenvector $\mathbf{u}_m$ associated with the eigenvalue ω_m ($m = 1$ or 2). The characteristic equation for the latter (eq. (4.5)) takes the explicit form

$$\omega_m^2 - T\omega_m + \Delta = 0 \qquad m = 1, 2 \qquad (4.17a)$$

where T and Δ are, respectively, the trace and the determinant of the matrix $\{\mathscr{L}_{ij}\}$,

$$\left.\begin{aligned} T &= \mathscr{L}_{11} + \mathscr{L}_{22} \\ \Delta &= \mathscr{L}_{11}\mathscr{L}_{22} - \mathscr{L}_{12}\mathscr{L}_{21} \end{aligned}\right\} \quad (4.17b)$$

The solution of the quadratic eq. (4.17a) is

$$\omega_{1,2} = \frac{T \pm (T^2 - 4\Delta)^{1/2}}{2} = \frac{T \pm \mathscr{D}^{1/2}}{2} \qquad (4.18)$$

where $\mathscr{D}$ is the discriminant. The nature of the roots depends on the signs of T, Δ and $\mathscr{D}$. The various distinct possibilities (Jordan and Smith, 1977; Andronov *et al.*, 1966; Cesari, 1963) are classified below, using the representation in the plane (Imω, Reω).

$\mathscr{D} > 0$: two real eigenvalues

(a) $\Delta > 0$, roots have the same sign

(a1) $T < 0$

(a2) $T > 0$

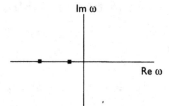

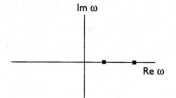

(b) $\Delta < 0$, roots have opposite signs

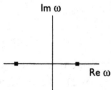

(b1) $T < 0$

(b2) $T > 0$

(b3) $T = 0$

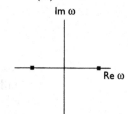

(c) $\Delta = 0$, at least one of the real roots is zero

(c1) $T < 0$

(c2) $T > 0$

(c3) $T = 0$

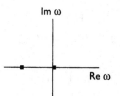

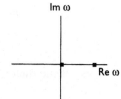

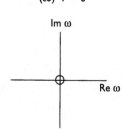

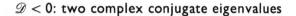

$\mathscr{D} < 0$: two complex conjugate eigenvalues

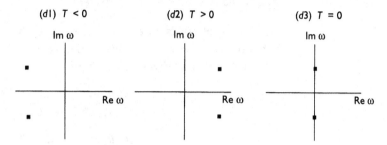

$\mathscr{D} = 0$: a double eigenvalue

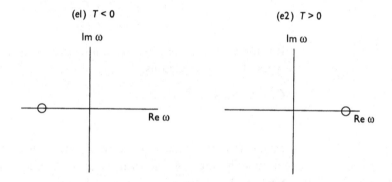

Cases ($b3$), ($c3$) and ($d3$) are the only ones that can be realized in conservative systems. Indeed, $T = \sum_i (\partial F_i/\partial X_i)_{X_s} = (\text{div}\mathbf{F})_{X_s}$, and by definition $\text{div}\mathbf{F} = 0$ in such systems. In a dissipative system these three cases along with ($c1$)–($c2$) and ($e1$)–($e2$) are nongeneric, in the sense that they require a strict equality which can only be realized (if at all) for specific values of the control parameter λ. They are important in the sense that they are borderline cases. In particular ($c1$) and ($d3$) define the borderline between asymptotic stability and instability.

Barring degeneracies we shall now consider more explicitly ($a1$)–($a2$), ($b1$)–($b2$) and ($d1$)–($d2$), our objective being to determine, in these generic cases, the shape of the phase portraits around the reference fixed point. We start by writing the solution of (4.15) in the explicit form (cf. eq. (4.6))

$$\left. \begin{aligned} x_1(t) &= C_1 u_{11}\, e^{\omega_1 t} + C_2 u_{21}\, e^{\omega_2 t} \\ x_2(t) &= C_1 u_{12}\, e^{\omega_1 t} + C_2 u_{22}\, e^{\omega_2 t} \end{aligned} \right\} \quad (4.19)$$

where the coefficients u_{ij} ($i, j = 1, 2$) are to be calculated from eqs. (4.16) and C_1, C_2 are fixed by the initial conditions. To go further we need to specify the type of situation considered.

(a) Two real roots of equal sign ($\mathcal{D} > 0$, $\Delta > 0$)

Suppose first $\omega_2 < \omega_1 < 0$. Differentiating the two relations (4.19) with respect to time and dividing we obtain

$$\frac{dx_2}{dx_1} = \frac{\omega_1 C_1 u_{12} e^{\omega_1 t} + \omega_2 C_2 u_{22} e^{\omega_2 t}}{\omega_1 C_1 u_{11} e^{\omega_1 t} + \omega_2 C_2 u_{21} e^{\omega_2 t}}$$

$$= \frac{\omega_1 C_1 u_{12} + \omega_2 C_2 u_{22} e^{(\omega_2 - \omega_1)t}}{\omega_1 C_1 u_{11} + \omega_2 C_2 u_{21} e^{(\omega_2 - \omega_1)t}} \qquad (4.20a)$$

Similarly, dividing the two eqs. (4.19) we obtain:

$$\frac{x_2}{x_1} = \frac{C_1 u_{12} + C_2 u_{22} e^{(\omega_2 - \omega_1)t}}{C_1 u_{11} + C_2 u_{21} e^{(\omega_2 - \omega_1)t}} \qquad (4.20b)$$

It is useful to consider first two limiting cases, corresponding to special types of initial conditions.

$C_1 = 0$. eq.(4.20b) reduces to

$$x_2/x_1 = u_{22}/u_{21} \qquad (4.21a)$$

In the (x_2, x_1) plane this is an equation of a straight line passing through the origin. As the ωs are negative, trajectories starting on this line tend to (0,0) in the course of time (Fig. 4.2)

Fig. 4.2 Two-dimensional phase portrait around an asymptotically stable node obtained from numerical integration of the system $dx_1/dt = x_1 - 2x_2$, $dx_2/dt = 3x_1 - 4x_2$. The trajectories come from infinity with a slope $u_{22}/u_{21} = \frac{3}{2}$, and converge to the fixed point with a slope tending to $u_{12}/u_{11} = 1$, eqs. (4.21a) and (4.21b).

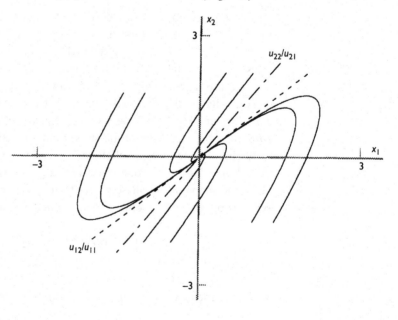

$C_2 = 0$. eq. (4.20b) reduces to

$$x_2/x_1 = u_{12}/u_{11} \qquad (4.21b)$$

which in the (x_2, x_1) plane represents another pair of straight line trajectories directed toward the origin (Fig. 4.2).

We consider now the general case $C_1 \neq 0$, $C_2 \neq 0$, corresponding to initial conditions on the plane outside the above two lines. When $t \to \infty$ the negative exponential in (4.20a) will tend to zero and the trajectories (no longer straight lines) will tend to the fixed point with a slope equal to that of the straight line (4.21b). In the opposite limit $t \to -\infty$ the exponential in (4.20a) explodes. The remaining terms can be neglected, and the trajectories come from infinity with a slope equal to that of the straight line (4.21a). We obtain in this way the configuration of Fig. 4.2. The corresponding fixed point will be referred to as a *node*. In Fig. 4.2 the node is asymptotically stable and is, therefore, the *attractor* of our dynamical system. Had we chosen ω_1 and ω_2 positive a similar topological configuration would have been realized, in which the trajectories would now diverge rather than converge. We then speak of an unstable node which can be referred to as a *repellor* (in contrast to the attracting node).

(b) Two real roots of opposite sign ($\mathscr{D} > 0, \Delta < 0$)

Suppose, without loss of generality, that $\omega_1 > 0$ and $\omega_2 < 0$. In the particular case $C_1 = 0$ the influence of the growing exponential disappears in (4.19) and the corresponding trajectories are straight lines given again by eq. (4.21a), approaching the origin. We refer to this pair of straight lines as the *stable manifold*, W_s of the fixed point. When $C_2 = 0$ only the unstable mode is excited in (4.19). Because of the cancellations taking place in eqs. (4.20) the trajectories are still straight lines, given now by (4.21b), but this time they are directed away from the fixed point. We shall refer to them as the *unstable manifold*, W_u of the fixed point. All other trajectories are hyperbola-like whose asymptotes as $t \to \infty$ and $t \to -\infty$ are, by the same argument as in the previous case, the two straight lines (4.21a) and (4.21b), as seen in Fig. 4.3. These lines divide the phase space into four regions between which passage is prohibited by virtue of the nonintersection of trajectories imposed by the theorem of uniqueness of solutions. They are referred to as *separatrices*, the fixed point itself being qualified as a *saddle point*. As the phase portrait of Fig. 4.3 shows, a saddle point combines a stabilizing action in one direction with a destabilizing one along another direction. Eventually the instability takes over unless the system is itself found initially on the line of eq. (4.21a). Since this line is of measure zero in the plane this situation is untypical: a saddle point in a

Fig. 4.3 Two-dimensional phase portrait around a saddle point obtained from numerical integration of the system
$dx_1/dt = 2x_1 + 2x_2$
$dx_2/dt = -2x_1 - 3x_2$.
The slopes of the stable and unstable manifolds (dotted lines) are, respectively,
$u_{22}/u_{21} = -2$ and $u_{12}/u_{11} = -\frac{1}{2}$.

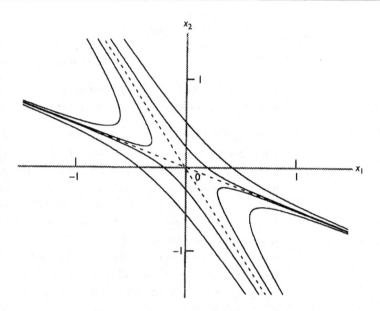

dynamical system is, therefore, a repellor. In a realistic system a runaway to infinity cannot take place. The separatrices will then bend and will either tend to new attracting sets or (in high-dimensional systems) be reinjected back to the vicinity of the fixed point. These phenomena are typically nonlinear and will be discussed further later in this monograph. Notice that saddle points can occur in conservative as well as in dissipative systems whereas, as stressed already above, nodes can only occur in dissipative systems.

(c) Two complex conjugate roots with a nonvanishing real part ($\mathscr{D} < 0, T \neq 0$).
Since $\omega_1 = \omega_2^*$, it follows from (4.16) that $u_{11} = u_{21}^*, u_{12} = u_{22}^*$. Since x_1 and x_2 are real, eq. (4.19) entails that $C_1 = C_2^*$. Setting

$$\omega_{1,2} = \mu \pm i\Omega \tag{4.22}$$

one may then write eqs. (4.19) as

$$\left.\begin{aligned}
x_1(t) &= e^{\mu t}(C_1 u_{11} e^{i\Omega t} + \text{cc}) \\
x_2(t) &= e^{\mu t}(C_1 u_{12} e^{i\Omega t} + \text{cc})
\end{aligned}\right\} \tag{4.23}$$

Introducing new amplitude and phase variables C, K, γ and κ through

$$\left.\begin{aligned}
C_1 u_{11} &= \tfrac{1}{2}C\, e^{i\gamma} \\
u_{12}/u_{11} &= K\, e^{i\kappa}
\end{aligned}\right\} \tag{4.24}$$

we may transform (4.23) into the more transparent form

$$x_1(t) = C\,e^{\mu t}\cos(\Omega t + \gamma) \left.\vphantom{\begin{array}{c}a\\b\end{array}}\right\} \quad (4.25)$$
$$x_2(t) = CK\,e^{\mu t}\cos(\Omega t + \kappa + \gamma)$$

These relations predict an oscillatory behavior around the fixed point. The oscillation is damped if $\mu < 0$ (i.e. $T < 0$), and amplified if $\mu > 0$ (i.e. $T > 0$). In the first case the fixed point is an (asymptotically stable) attractor, in the second an (unstable) repellor. Notice that while in eqs. (4.24) and (4.25) C and γ are determined by the initial conditions, the parameters K and κ are intrinsic, as they are determined by the linearized problem.

To obtain the phase space portrait around the fixed point we switch to the canonical representation in which the matrix $\mathcal{L}$ is diagonalized (cf. eqs. (4.7)–(4.9)),

$$dz/dt = (\mu + i\Omega)z \quad (4.26)$$

Only one such relation is needed in the present two-variable system, since the second equation featuring the eigenvalue $\mu - i\Omega$ would merely be the complex conjugate of (4.26). Introducing the polar coordinates r, ϕ through

$$z = r\,e^{i\phi} \quad (4.27)$$

and separating real and imaginary parts we further transform (4.26) into

$$dr/dt = \mu r \left.\vphantom{\begin{array}{c}a\\b\end{array}}\right\} \quad (4.28)$$
$$d\phi/dt = \Omega$$

or, eliminating the time between the two equations,

$$\frac{dr}{d\phi} = \left(\frac{\mu}{\Omega}\right) r \quad (4.29a)$$

or finally

$$r = r_0\,e^{[(\mu/\Omega)\phi]} \quad (4.29b)$$

where r_0 is an integration constant. This equation represents a family of logarithmic spirals. If $\mu < 0$ the representative point will tend to the origin (Fig. 4.4); otherwise the trajectories will spiral away from the origin. In both cases the fixed point is referred to as the *focus*. The domains of asymptotic stability and instability are separated by the critical condition $T = 0$, or $\mu = 0$. In this case eqs. (4.28) reduce to a family of circles surrounding the origin, which is referred to as a *center* (Fig. 4.5). As pointed out earlier an attracting focus can only arise in dissipative systems, whereas a center is compatible with a conservative (e.g.

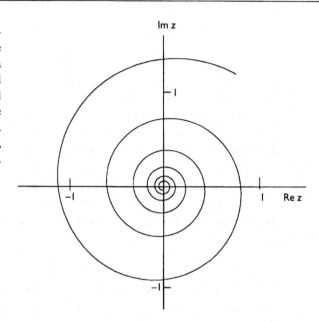

Fig. 4.4 Two-dimensional phase portrait around a stable focus obtained from numerical integration of the canonical form of eq. (4.26) with $\mu = -0.2$, $\Omega = 1$.

Hamiltonian) system. In a dissipative system the center is nongeneric, in the sense that it is an exceptional situation in which the parameters present in the problem must satisfy a strict equality. This is further reflected by the fact that under the slightest change of parameters the topology of the phase portrait will undergo a *qualitative* change, switching from the form of Fig. 4.5 to that of Fig. 4.4. We refer to this phenomenon as *structural instability*. In the same sense a node, a saddle point or a focus arising in a dissipative system is structurally stable. Hamiltonian (more generally conservative) dynamical systems, although nongeneric and structurally unstable in the abstract space of all possible dynamical systems, are nevertheless generic from the standpoint of physics. They remain robust under a special class of transformations, the only ones that seem to be allowed by the laws of nature, which we already referred to in Section 2.1 as *canonical transformations*.

4.4 Examples of stability analysis of two-dimensional dynamical systems

Following the order of presentation of Chapter 2 we first consider the problem of the hoop as a typical example of a two-dimensional conservative dynamical system. The equation of evolution of this system was derived in Section 2.1, eq. (2.8a). For the purposes of the stability

Fig. 4.5 Phase portrait when the limit $\mu = 0$ is taken in eq. (4.26) and Fig. 4.4. The fixed point behaves as a center.

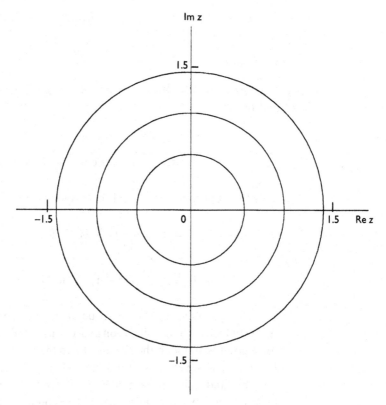

analysis it will be convenient to transform this second order equation into a pair of first order ones:

$$
\left.
\begin{aligned}
\mathrm{d}\theta/\mathrm{d}t &= v \\
\mathrm{d}v/\mathrm{d}t &= (g/r)\sin\theta(\lambda\cos\theta - 1)
\end{aligned}
\right\} \quad (4.30)
$$

According to the general procedure laid down in the preceding sections, the first step is to determine the fixed points $(\theta_s, 0)$, where θ_s is given by $\sin\theta_s(\lambda\cos\theta_s - 1) = 0$. In the domain of variation of θ, $-\pi \leq \theta \leq \pi$, the solutions to this equation are

$$\theta_{s0} = 0 \qquad (4.31a)$$

$$\theta_{s,\pm} = \arccos\lambda^{-1} \qquad (4.31b)$$

The first solution exists for all parameter values, but for the second one the condition $\lambda > 1$ is required. Referring to (2.8b), this means that the hoop must rotate with an angular velocity ω such that (cf. also Fig. 1.1)

$$\omega \geq \omega_c = (g/r)^{1/2} \qquad (4.32)$$

The next step is to introduce perturbations around the fixed points,

$$\left.\begin{array}{l} \theta = \theta_s + \delta\theta \\ v = \delta v \end{array}\right\} \quad (4.33)$$

Linearizing eqs. (4.30) with respect to $\delta\theta$ and δv we obtain, using (3.27a) and (4.15),

$$\left.\begin{array}{l} \mathrm{d}\delta\theta/\mathrm{d}t = \delta v \\ \mathrm{d}\delta v/\mathrm{d}t = \dfrac{g}{r}[\cos\theta_s(\lambda\cos\theta_s - 1) - \lambda\sin^2\theta_s]\,\delta\theta \end{array}\right\} \quad (4.34)$$

In the notation of (4.17) this corresponds to

$$T = 0, \Delta_0 = \frac{g}{r}(1 - \lambda) \qquad \text{for } \theta_s = \theta_{s0} = 0 \qquad (4.35\text{a})$$

$$T = 0, \Delta_\pm = \frac{g}{\lambda r}(\lambda^2 - 1) \qquad \text{for } \theta_s = \theta_{s\pm} \text{ and } \lambda \geqslant 1 \quad (4.35\text{b})$$

For $\lambda < 1$ only (4.35a) has to be taken into consideration. One has $\Delta_0 > 0$ which in the classification of Section 4.3 falls in case (d3): the trivial fixed point is stable in the sense of Lyapunov and behaves like a center. For $\lambda > 1$, however, $\Delta_0 < 0$ and case (b3) applies: the trivial fixed point is unstable and behaves as a saddle. Under the same conditions $\Delta_\pm$ in (4.35b) is positive and the nontrivial fixed points behave as a center. This *exchange of stability* is in full agreement with Fig. 1.1 and the physics of the problem as discussed in Sections 1.2 and 2.1. Notice that in the presence of friction the situation would change in a qualitative way. Lyapunov stability would be replaced by asymptotic stability and the center would unfold to become a focus.

We now turn to dissipative systems considering the Brusselator model (eqs. (2.44)) as a representative example. The linearized equations around the unique fixed point $(A, B/A)$ have already been written down in (3.28),

$$\mathrm{d}x/\mathrm{d}t = (B - 1)x + A^2 y$$

$$\mathrm{d}y/\mathrm{d}t = -Bx - A^2 y$$

and the characteristic equation (eq. (4.17a)) reads

$$\omega^2 - (B - 1 - A^2)\omega + A^2 = 0 \qquad (4.36)$$

Since in the notation of Section 3.3 $\Delta > 0$, the fixed point can never be a saddle. The discriminant

$$\mathscr{D} = (B - 1 - A^2)^2 - 4A^2$$

can be further written as

$$\mathscr{D} = [B - (A + 1)^2][B - (A - 1)^2] \qquad (4.37a)$$

whereas the trace T is

$$T = B - 1 - A^2$$

As the parameter B varies while the second parameter A is kept fixed, the fixed point switches from asymptotic stability to instability at $T = 0$, or

$$B_c = A^2 + 1 \qquad (4.38)$$

It can behave as node or focus, depending on the value of B relative to $(A - 1)^2$ or $(A + 1)^2$, as summarized in the diagram below:

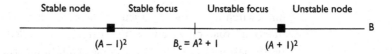

4.5 Three variables and beyond

In a system involving three variables the characteristic equation (eq. (4.5)) around a fixed point will be an algebraic equation of third degree with real coefficients, which can have either three real roots or one real and two complex conjugate roots. Fig 4.6(a)–(e) depicts the various (generic) possibilities for the positions of these roots in the complex plane, it being understood that to each of these cases correspond cases (a')–(e') obtained by reflection with respect to the imaginary axis. There exist, in addition,

Fig. 4.6 The five possible (nondegenerate) configurations of the roots of the characteristic equation of a three-dimensional dynamical system in the complex plane.

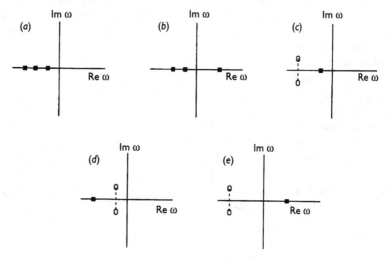

degenerate cases whereby the real part of at least one of the roots is zero, the real parts of two roots (not belonging to a complex conjugate pair) are equal, and combinations thereof. As stressed in Sections 4.2 and 4.3 these are borderline cases signaling the passage from asymptotic stability to instability. They are important in their own right, but since they can be handled along similar lines as before we do not consider them in this section and focus, instead, on some of the new possibilities arising from the presence of an additional variable. In this perspective cases (*a*) and (*b*) are straightforward generalizations of the node and saddle of the two-dimensional case, and need not be considered further.

Cases (*c*) and (*d*) are new, since they are combinations of situations corresponding to a node and to a focus. Writing the solution in the form

$$\mathbf{x} = C_1 \mathbf{u}_1\, e^{\omega_1 t} + C_2 \mathbf{u}_2\, e^{\omega_2 t} + C_3 \mathbf{u}_3\, e^{\omega_3 t} \qquad (4.39)$$

and switching to the representation in which the linearized operator is diagonal one may reduce the original (coupled) equations for the perturbations x_1, x_2, x_3 to a complex equation corresponding to the pair of the complex eigenvalues ω_1, ω_2 and a real one corresponding to the third eigenvalue ω_3,

$$\left.\begin{aligned} dz/dt &= (\mu + i\Omega)z \\ d\zeta/dt &= \omega_3 \zeta \end{aligned}\right\} \qquad (4.40)$$

This entails that the fixed point behaves like a (stable) focus in the subspace spanned by the eigenvectors $\mathbf{u}_1, \mathbf{u}_2$, while along the third direction $\mathbf{u}_3$ it is approached monotonically at a rate equal to ω_3. The corresponding three-dimensional phase portrait is depicted in Fig. 4.7. The trajectories are directed toward the fixed point following a funnel converging to this point (*a*), or a paraboloid of revolution (*b*), according to whether $\mathrm{Re}\,\omega_{1,2} < \omega_3 < 0$ or $\omega_3 < \mathrm{Re}\,\omega_{1,2} < 0$ (Arnol'd, 1980).

Fig. 4.7 Three-dimensional phase portrait around a fixed point obtained from numerical integration of the canonical form of eq. (4.40): (*a*) $\mu = -0.15$, $\omega_3 = -0.05$, $\Omega = 1$ (case (*c*) of Fig. 4.6); (*b*) $\mu = -0.05$, $\omega_3 = -0.15$, $\Omega = 1$ (case (*d*) of Fig. 4.6).

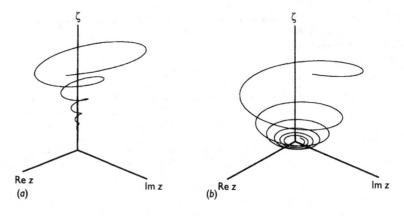

(*a*)

(*b*)

Case (*e*) is by far the most interesting. In the same representation as above, the fixed point behaves in the subspace spanned by $\mathbf{u}_1$, $\mathbf{u}_2$ as a stable focus, while along the third direction $\mathbf{u}_3$ the trajectory evolves away from this point. In the three-dimensional phase space the trajectories evolve away from the fixed point following an inverted funnel (Fig. 4.8). This configuration, which is reminiscent of both a saddle and a focus, is referred to as a *saddle-focus*. In the case (*e*) of Fig. 4.6 the corresponding fixed point possesses a two-dimensional *stable manifold* ($\mathbf{u}_1$, $\mathbf{u}_2$) and a one-dimensional *unstable manifold* ($\mathbf{u}_3$), but in the symmetric case (*e'*) the stable manifold would be one-dimensional and the unstable one two-dimensional.

The importance of the saddle-focus in dynamical systems stems from the fact that it combines in a single dynamics a stabilizing trend coexisting with a destabilizing one, while allowing at the same time for oscillatory behavior. In the linearized case the trajectories will inevitably tend to infinity, but in the original nonlinear problem the exclusion of runaway effects will force the stable and unstable manifolds and, consequently, the trajectories themselves to bend and to remain confined. In the absence of an attractor in the form of another fixed point or a limit cycle this may result in a very intricate motion consisting of an aperiodic succession of unstable stages removing the trajectory from the fixed point, followed by a *reinjection* back to the vicinity of the fixed point. It can be shown (Shil'nikov, 1965) that this will actually be the case if the parameters are such that the system operates near a situation in which the stable and unstable manifolds merge. This possibility, to which we have alluded already briefly in Section 3.3B, implies the existence of *homoclinic*

Fig. 4.8 Three-dimensional portrait around a saddle-focus (case (*e*) of Fig. 4.6), obtained from numerical integration of eqs. (4.40) with $\mu = -0.15$, $\omega_3 = 0.05$, $\Omega = 1$.

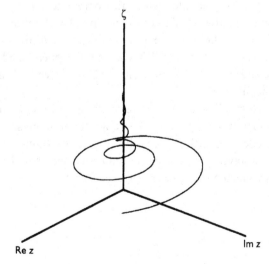

Re z Im z

trajectories. Such trajectories are structurally unstable since the merging condition requires that certain equalities between parameters be fulfilled. But the important point is that, if some inequalities depending on the eigenvalues of the saddle-focus known as Shil'nikov conditions are fulfilled, the destruction of a homoclinic trajectory will give rise to a rich structure in phase space allowing for deterministic chaos. For this reason, the saddle-focus and the associated homoclinic orbit can be regarded as important organizing centers for chaos in three- and higher-dimensional dynamical systems. We stress that the flexibility allowed by the third dimension is essential, since in a two-dimensional space a sequence of bendings and reinjections would be impossible without intersection of the trajectories, which is prohibited by the uniqueness theorem of the solutions of eqs. (3.6).

Homoclinic behavior has been found in the Lorenz model introduced in Section 3.1 (Sparrow, 1982; Glendinning and Sparrow, 1984). An elegant model showing the role of homoclinicity in the onset of chaos has been introduced by Rössler (1976, 1979). It consists of three coupled equations with a single (quadratic) nonlinearity,

$$\left. \begin{aligned} \mathrm{d}x/\mathrm{d}t &= -y - z \\ \mathrm{d}y/\mathrm{d}t &= x + ay \\ \mathrm{d}z/\mathrm{d}t &= bx - cz + xz \end{aligned} \right\} \qquad (4.41)$$

The characteristic equation around the fixed point $x = y = z = 0$ is

$$\omega^3 + (c - a)\omega^2 + (1 + b - ac)\omega + c - ab = 0 \qquad (4.42)$$

and may generate, when the parameters are varied, a saddle-focus satisfying the Shil'nikov condition. Furthermore when $a = 0.38$, $b = 0.30$, $c = 4.82$ there exists a homoclinic orbit, shown in Fig. 4.9. If c decreases toward $c = 4.5$ this orbit disappears and a chaotic attractor is generated, the structure of which is shown in Fig. 4.10 (Gaspard and Nicolis, 1983).

Homoclinicity as a route to chaos arises in higher-than-three-dimensional systems as well. There are some interesting new phenomena in connection with the possibility that the stable and unstable manifolds themselves can now be high-dimensional, but their detailed analysis is beyond the scope of the present monograph.

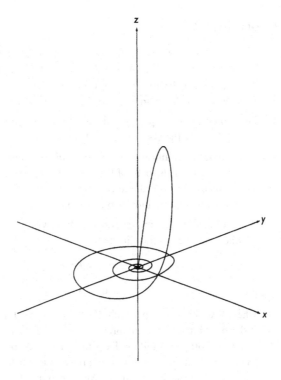

Fig. 4.9 Homoclinic orbit associated with the fixed point $(0, 0, 0)$ of Rössler's model, eqs. (4.41), for parameter values $a = 0.38$, $b = 0.30$, $c = 4.82$.

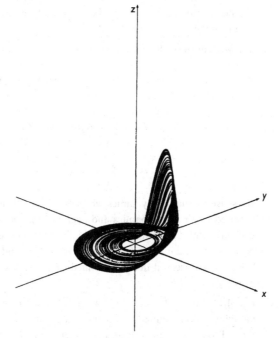

Fig. 4.10 Chaotic attractor obtained from numerical integration of Rössler's model, eqs. (4.41), for $a = 0.32$, $b = 0.30$, $c = 4.50$. The trajectories are injected on the same side of the unstable fixed point, a situation referred to as *spiral chaos*.

Problems

4.1 Show that the Semenov equation (eq. (2.47)) can admit up to three steady
state solutions and determine the linear stability of these solutions using
the cooling coefficient α as a control parameter. What happens when
consumption of reactants is allowed (cf. also Problem 2.11)?

4.2 Determine the fixed points and their linear stability of the Brusselator in
a CSTR under the conditions of Problem 2.6.

4.3 Derive eq. (4.42) and find the domain of parameter values for which the
trivial fixed point $(0,0,0)$ of the Rössler model (eq. (4.41)) behaves as a
saddle-focus. Analyze the linear stability of the nontrivial fixed point of
the model (Gaspard and Nicolis, 1983).

4.4 The Volterra–Lotka model (Lotka, 1924; Volterra, 1936) describes the
predator–prey dynamics in the form

$$dx/dt = kx - sxy$$

$$dy/dt = sxy - fy$$

where x and y are respectively the prey and predator population densities,
k the prey birth rate, f the predator death rate and s the frequency of
predator–prey encounters. Determine the fixed points of this system and
their linear stability (a) if k is regarded as a constant, (b) if a regulation in
the sense of Verhulst (Section 2.6) is introduced, $k = a - bx$. Show that
under the change of variables $u = 1nx$, $v = 1ny$ the equations are
transformed (in case (a)) to a Hamiltonian form and identify the effective
Hamiltonian (Kerner, 1957).

4.5 The Willamowski–Rössler model

$$A_1 + X \underset{k_{-1}}{\overset{k_1}{\rightleftharpoons}} 2X \qquad X + Y \underset{k_{-2}}{\overset{k_2}{\rightleftharpoons}} 2Y$$

$$A_5 + Y \underset{k_{-3}}{\overset{k_3}{\rightleftharpoons}} A_2 \qquad X + Z \underset{k_{-4}}{\overset{k_4}{\rightleftharpoons}} A_3$$

$$A_4 + Z \underset{k_{-5}}{\overset{k_5}{\rightleftharpoons}} 2Z$$

gives rise to chaotic dynamics while satisfying all the requirements
imposed by thermodynamics and chemical kinetics (Willamowski and
Rössler, 1980; Geysermans and Nicolis, 1993). Determine the fixed points
of the rate equations and the parameter values under which saddle-focus
behavior is observed in the simplified case $k_{-2} = k_{-3} = k_{-4} = 0$,
$k_2 = k_4 = 1$.

4.6 Fig. 4.11 represents (left) a voltaic arc connected in series with an
inductance and shunted by a capacitance and (right) the dependence of
the voltage across the arc on the current i. (a) Derive the equations for

the current i and voltage u across the capacitance. (*b*) Determine the fixed points and their linear stability using the slope of the $\psi(i)$ curve at these points as a control parameter. (*c*) Study the cases $C = 0$, $L \neq 0$ and $C \neq 0$, $L = 0$ and compare the results with the ones obtained in (*b*) in the limit of very small C or L (Andronov *et al.*, 1966).

4.7 The global energy balance of the planet earth is described qualitatively by the equation

$$C(\mathrm{d}T/\mathrm{d}t) = (\text{incoming solar energy}) - (\text{outgoing infrared energy})$$

$$= Q[1 - \alpha(T)] - \varepsilon\sigma T^4$$

where T is the space averaged surface temperature, C the heat capacity, Q the solar constant, α the albedo, σ the Stefan constant and ε an emissivity factor accounting for deviations from black body radiation (Crafoord and Källén, 1978). By modeling the albedo as a piecewise linear function of T with two extreme horizontal branches and an intermediate one of negative slope, show that this equation may admit three steady state solutions, two of which are stable and one unstable. Discuss quantitatively the case:

$$\alpha = 0.8 \qquad\qquad T < T_1$$

$$\alpha = 0.25 \qquad\qquad T > T_2$$

$$\alpha = a - bT \qquad\quad T_1 < T < T_2, a = 2.75, b = 0.0085$$

Parameter values: $Q = 340\mathrm{W\,m^{-2}}$, $\varepsilon = 0.61$.

4.8 A qualitative description of the coupling between mean ocean temperature and sea ice extent is provided by the system of equations (Saltzman, Sutera and Hansen, 1982; Nicolis, 1984)

$$\mathrm{d}\eta/\mathrm{d}t = \theta - \eta$$

$$\mathrm{d}\theta/\mathrm{d}t = b\theta - a\eta - \eta^2\theta$$

where η and θ are, respectively, suitably scaled deviations of the latitude of sea ice extent and of the mean ocean temperature from a reference state. Compute the fixed points of this system and determine their type and stability in terms of the parameters a and b.

Fig.4.11

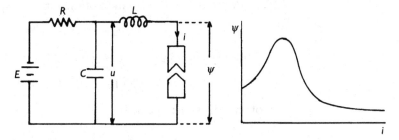

CHAPTER FIVE

Nonlinear behavior around fixed points: bifurcation analysis

5.1 Introduction

The importance of linear stability analysis is to show that a qualitative change of behavior may occur within a single, well-defined dynamical system beyond the critical value λ_c of the control parameter at which the system switches from asymptotic stability to instability. However, as soon as one enters the domain of instability the linearized equations become inadequate, as they predict runaway to infinity. In order to investigate the existence of new physically acceptable solutions which emerge beyond the threshold of instability the full, nonlinear equations will have to be analyzed. This is the objective of the present chapter.

The starting point is given by eqs. (3.26),

$$\mathrm{d}\mathbf{x}/\mathrm{d}t = \mathscr{L}(\lambda) \cdot \mathbf{x} + \mathbf{h}(\mathbf{x}, \lambda) \tag{5.1}$$

We suppose that linear stability analysis performed on these equations along the lines of the previous chapter has established the existence of a critical value λ_c such that the linearized operator $\mathscr{L}(\lambda_c)$ admits an eigenvalue with vanishing real part, $\mathrm{Re}\,\omega_c = \mathrm{Re}\,\omega(\lambda_c) = 0$. The linearized version of (5.1),

$$\mathrm{d}\mathbf{x}/\mathrm{d}t = \mathscr{L}(\lambda) \cdot \mathbf{x} \tag{5.2}$$

then admits at $\lambda = \lambda_c$ a solution of the form (cf.eq.(4.1))

$$\mathbf{x} = \mathbf{u}\, e^{i(\mathrm{Im}\,\omega_c)t} = \mathbf{u}\, e^{i\Omega_c t} \tag{5.3}$$

Substituting into (5.2) and setting $\lambda = \lambda_c$ one finds that

$$[i\Omega_c \mathbf{1} - \mathscr{L}(\lambda_c)] \cdot \mathbf{u} = 0 \tag{5.4}$$

in other words, the operator

$$\mathbf{J}_c = i\Omega_c \mathbf{1} - \mathscr{L}(\lambda_c) \tag{5.5}$$

admits at least one eigenvector **u** corresponding to a zero eigenvalue. We also express this property by the statement that J_c admits a *nontrivial null space*. The question is now, what is the behavior of the solutions of the full nonlinear problem (eqs. (5.1)) for values of the control parameter λ in a certain neighborhood of λ_c. The following two theorems give a surprisingly comprehensive answer.

Theorem 1   (Sattinger, 1972). If

x = 0 remains a solution of (5.1) in a neighborhood of λ_c,
ω_c is a simple eigenvalue that is a simple root of the characteristic equation (or more generally a root of odd multiplicity)
then $\lambda = \lambda_c$ is a *bifurcation point*, in the sense that there is at least one new branch of solutions outgoing from (**x** = 0, λ_c). This branch either extends to infinity or meets another bifurcation point.

Theorem 2   If ω_c is a simple eigenvalue and the additional *transversality condition* is satisfied,

$$\left[\frac{d}{d\lambda}\operatorname{Re}\omega(\lambda)\right]_{\lambda=\lambda_c} \neq 0 \tag{5.6}$$

guaranteeing that the Reω versus λ curve in Fig. 4.1 crosses the λ – axis at $\lambda = \lambda_c$, then

the bifurcating solutions will be stationary if $\Omega_c = 0$ in (5.4);
the bifurcating solutions will be time-periodic if $\Omega_c \neq 0$ in (5.4) (*Hopf bifurcation*);
in both of the above cases *supercritical* branches (bifurcating in the region of λ-values for which the reference state has lost its stability) are stable and *subcritical* ones (bifurcating in the region of λ-values for which the reference state is stable) are unstable, provided that the remaining eigenvalues of $\mathscr{L}(\lambda_c)$ have negative real parts.

Fig. 5.1 summarizes the various possibilities

Rather than reproduce here one of the proofs of these theorems found in the mathematical literature we shall adopt a physicist's *constructive* approach in which the validity of the theorems will be verified by the analytic construction of the solutions of the full nonlinear problem. This approach is outlined in the subsequent sections.

5.2 Expansion of the solutions in perturbation series: the case of zero eigenvalue, Re ω_c = Im ω_c = 0

We have repeatedly stressed the difficulties arising in the solution of nonlinear problems. We therefore give up the idea of obtaining exact results of a global character, and limit our attention to the *local behavior* of the solutions in the vicinity of the bifurcation point λ_c. Furthermore, we suppose that the new solutions emerge at λ_c in a continuous fashion, thus excluding vertical branchings or jumps.

These hypotheses allow us to expand x in the vicinity of λ_c in power series of a small parameter. The latter must certainly be related to $\lambda - \lambda_c$, since at $\lambda = \lambda_c$ the norm $|x|$ of the solution goes to zero. There is no reason, however, for this small parameter to be $\lambda - \lambda_c$ itself, since in principle nothing guarantees the analyticity of the solutions in $\lambda - \lambda_c$. Later on in this chapter we shall in fact encounter several examples of manifestly nonanalytic dependence.

Fig. 5.1 The three elementary bifurcations at a simple eigenvalue arising under the conditions of theorems 1 and 2: full and dotted lines represent, respectively, asymptotically stable (S) and unstable (U) branches of solutions. In (a) and (b) the amplitude of the solution is plotted versus the control parameter; (c) schematically depicts, in addition, the continuous family of solutions corresponding to the different values of the phase of the oscillatory motion.

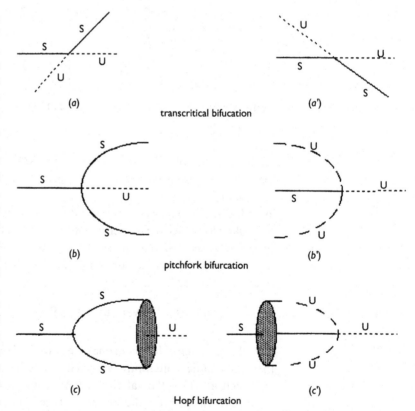

To allow the system itself to fix the dependence of the solutions on $\lambda - \lambda_c$ we introduce an auxiliary smallness parameter ε with respect to which we expand both $\mathbf{x}$ and $\lambda - \lambda_c$:

$$\mathbf{x} = \varepsilon \mathbf{x}_1 + \varepsilon^2 \mathbf{x}_2 + \cdots \tag{5.7}$$

$$\lambda - \lambda_c = \varepsilon \gamma_1 + \varepsilon^2 \gamma_2 + \cdots \tag{5.8}$$

where the coefficients $\gamma_1, \gamma_2, \ldots$ are to be determined from the perturbation analysis.

Eqs. (5.1) feature not only $\mathbf{x}$ but also its rate of change with respect to the independent variable t. At criticality, and for the case of $\Omega_c = 0$ considered in this section, eq. (5.3) entails that at the level of a linearized description $\mathbf{x}$ does not vary at all with time after a transient period during which the stable modes (eigenvalues of $\mathscr{L}(\lambda_c)$ with negative real parts) have relaxed exponentially to zero. By continuity, one expects that for λ close to λ_c the solution $\mathbf{x}$ of the full system of eqs. (5.1) will be a slowly varying function of time. This *critical slowing down*, reminiscent of the theory of equilibrium critical phenomena (Stanley, 1971) suggests the introduction of new, more relevant time scales τ_1, τ_2 etc. through

$$\frac{\mathrm{d}}{\mathrm{d}t} = \varepsilon \frac{\partial}{\partial \tau_1} + \varepsilon^2 \frac{\partial}{\partial \tau_2} + \cdots \tag{5.9}$$

We shall refer to the scheme defined by eqs. (5.7)–(5.9) as *multiscale perturbation expansion* (Keworkian and Cole, 1981). Under certain conditions met in typical applications the convergence of this scheme can be guaranteed, but the radius of convergence can, in general, not be determined.

Substituting (5.7)–(5.9) into eqs. (5.1) we get, to the various orders in ε, the following systems of equations:

A $\mathcal{O}(\varepsilon)$

We obtain a single contribution,

$$\mathscr{L}(\lambda_c) \cdot \mathbf{x}_1 = 0 \tag{5.10}$$

B $\mathcal{O}(\varepsilon^2)$

We obtain four kinds of contribution. First, the operator $\mathscr{L}(\lambda_c)$ may act on the second order term $\mathbf{x}_2$ in the expansion of $\mathbf{x}$, eq. (5.7). Second, we may evaluate the operator $\mathscr{L}(\lambda)$ at a λ close to λ_c, and have it act on $\mathbf{x}_1$. This shift will be expressed formally by the first term of a Taylor expansion of $\mathscr{L}(\lambda)$ around λ_c and will thus yield the contribution $\gamma_1 \mathscr{L}_\lambda(\lambda_c) \cdot \mathbf{x}_1$, where derivatives are denoted by subscripts. Third, to order ε^2 we will have a contribution coming from the quadratic part of $\mathbf{h}$ in eqs. (5.1) which will be denoted formally as the second order term of a Taylor expansion around zero. Finally, to this order we will have a first contribution of the time derivative in the left hand side of (5.1) featuring the new time scale τ_1. We thus obtain the full equation to order ε^2:

$$\mathscr{L}(\lambda_c)\cdot\mathbf{x}_2 = -\gamma_1\mathscr{L}_\lambda(\lambda_c)\cdot\mathbf{x}_1 - \frac{1}{2}\mathbf{h}_{xx}\cdot\mathbf{x}_1\mathbf{x}_1 + \partial\mathbf{x}_1/\partial\tau_1 \qquad (5.11)$$

$$= \mathbf{q}_2$$

An example of explicit form of the various operators appearing in this equation is given in Appendix A2.

C $O(\varepsilon^3)$

Proceeding in the same manner as above we obtain

$$\mathscr{L}(\lambda_c)\cdot\mathbf{x}_3 = -\gamma_1\mathscr{L}_\lambda(\lambda_c)\cdot\mathbf{x}_2 - \gamma_2\mathscr{L}_\lambda(\lambda_c)\cdot\mathbf{x}_1 - \tfrac{1}{2}\gamma_1^2\mathscr{L}_{\lambda\lambda}(\lambda_c)\cdot\mathbf{x}_1$$

$$- \tfrac{1}{2}\gamma_1\mathbf{h}_{xx\lambda}(\lambda_c)\cdot\mathbf{x}_1\mathbf{x}_1 - \tfrac{1}{6}\mathbf{h}_{xxx}(\lambda_c)\cdot\mathbf{x}_1\mathbf{x}_1\mathbf{x}_1$$

$$- \mathbf{h}_{xx}(\lambda_c)\cdot\mathbf{x}_1\mathbf{x}_2 + \frac{\partial\mathbf{x}_1}{\partial\tau_2} + \frac{\partial\mathbf{x}_2}{\partial\tau_1} = \mathbf{q}_3 \qquad (5.12)$$

Similar expressions can be written to an arbitrary order in ε. The initial nonlinear problem has thus been replaced by an infinite sequence of *linear* problems. However, as we shall see shortly, eqs. (5.10)–(5.12) already contain the essential information needed for understanding bifurcation.

5.3 The amplitude equation: transcritical bifurcation

We want now to construct the solutions of eqs. (5.10)–(5.12). We notice that to $O(\varepsilon)$ the structure of eqs. (5.10) is identical to the linear stability problem (eq. (3.29)) at the criticality point $\lambda = \lambda_c$. The solution takes therefore the form (cf. eq. (4.1) for $\omega = \omega_c = 0$)

$$\mathbf{x}_1 = c(\tau_1, \tau_2, \cdots)\mathbf{u} \qquad (5.13)$$

We have factored out the amplitude c of the solution, which is undetermined at this stage in view of the fact that eqs. (5.10) are homogeneous. It is therefore understood from now on that $\mathbf{u}$ is completely determined from the linearized problem. For instance, in a two-variable system we have from (4.16):

$$\mathscr{L}_{11}(\lambda_c)u_1 + \mathscr{L}_{12}(\lambda_c)u_2 = 0$$

$$\mathscr{L}_{21}(\lambda_c)u_1 + \mathscr{L}_{22}(\lambda_c)u_2 = 0$$

with $\mathscr{L}_{11}(\lambda_c)\mathscr{L}_{22}(\lambda_c) - \mathscr{L}_{12}(\lambda_c)\mathscr{L}_{21}(\lambda_c) = 0$ from which $\mathbf{u}$ can be computed as

$$\mathbf{u} = \begin{pmatrix} 1 \\ -\dfrac{\mathscr{L}_{11}(\lambda_c)}{\mathscr{L}_{12}(\lambda_c)} \end{pmatrix} \tag{5.14}$$

Furthermore, since the slow time does not appear explicitly in eqs. (5.10) we have allowed in (5.13), for a dependence of c on the various slow time scales introduced in the perturbative expansion, eq. (5.9).

We next turn to $O(\varepsilon^2)$. Inserting (5.13) into (5.11) we obtain

$$\mathscr{L}(\lambda_c)\cdot\mathbf{x}_2 = -c\gamma_1\mathscr{L}_\lambda(\lambda_c)\cdot\mathbf{u} - \tfrac{1}{2}c^2\mathbf{h}_{xx}(\lambda_c)\cdot\mathbf{uu} + \frac{\partial c}{\partial\tau_1}\mathbf{u} = \mathbf{q}_2(\mathbf{u}, c, \lambda_c) \tag{5.15}$$

These relations constitute an *inhomogeneous* set of equations for $\mathbf{x}_2$, since the right hand side depends on the solution of the equations of the lower order. One is tempted to write, formally, its solution as

$$\mathbf{x}_2 \sim \mathscr{L}^{-1}(\lambda_c)\cdot\mathbf{q}_2 \tag{5.16}$$

The point, however, is that $\mathscr{L}(\lambda_c)$ is *not* everywhere invertible since by (5.4) (and $\Omega_c = 0$) it possesses a nontrivial null space. In such a case the action of the inverse operator according to (5.16) will produce divergent results, unless the parts of $\mathbf{q}_2$ responsible for the divergence can be eliminated. An important result of analysis, known as the theorem of the Fredholm alternative (Sattinger, 1972; Iooss and Joseph, 1980) prescribes how this can be achieved. To formulate the theorem one has to endow the space with a scalar product and define the adjoint operator $\mathscr{L}^+(\lambda_c)$. In the present case of dynamical systems involving a finite number of variables these can simply follow the conventional definitions of scalar product and adjoint of a matrix familiar from linear algebra (Friedman, 1956). For instance, taking once again the example of a two-variable system,

$$\mathscr{L}^+(\lambda_c) = \begin{pmatrix} \mathscr{L}_{11}(\lambda_c) & \mathscr{L}_{21}(\lambda_c) \\ \mathscr{L}_{12}(\lambda_c) & \mathscr{L}_{22}(\lambda_c) \end{pmatrix}$$

Let us introduce the null eigenvector $\mathbf{u}^+$ of $\mathscr{L}^+$ through

$$\mathscr{L}^+(\lambda_c)\cdot\mathbf{u}^+ = 0 \tag{5.17}$$

The theorem of the Fredholm alternative stipulates, then, that the right hand side $\mathbf{q}_2$ of (5.15) should be orthogonal (with the choice of scalar product discussed above) to the null eigenspace of $\mathscr{L}^+(\lambda_c)$,

$$(\mathbf{u}^+, \mathbf{q}_2) = 0 \tag{5.18}$$

This relation, also referred to as the *solvability condition*, follows straightforwardly by taking the scalar product of both sides of eq. (5.15) with $\mathbf{u}^+$ and by applying subsequently the definition of the adjoint operator. The point is that just like $\mathbf{u}$, the vector $\mathbf{u}^+$ is completely determined by (5.17). Furthermore it is unique, since we consider bifurcation at a simple eigenvalue. Eq. (5.18) reduces, therefore, to a single equation for the undetermined amplitude c. To find its explicit structure we substitute $\mathbf{q}_2$ as given by (5.15):

$$-c\gamma_1(\mathbf{u}^+, \mathscr{L}_\lambda(\lambda_c)\cdot\mathbf{u}) - \tfrac{1}{2}c^2(\mathbf{u}^+, \mathbf{h}_{\mathbf{xx}}(\lambda_c)\cdot\mathbf{uu}) + \frac{\partial c}{\partial\tau_1}(\mathbf{u}^+, \mathbf{u}) = 0$$

or

$$\partial c/\partial\tau_1 = \gamma_1 P_1 c - P_2 c^2 \tag{5.19}$$

where the coefficients P_1 and P_2 are given by

$$P_1 = \frac{1}{(\mathbf{u}^+, \mathbf{u})}(\mathbf{u}^+, \mathscr{L}_\lambda(\lambda_c)\cdot\mathbf{u})$$

$$P_2 = -\frac{1}{2(\mathbf{u}^+, \mathbf{u})}(\mathbf{u}^+, \mathbf{h}_{\mathbf{xx}}(\lambda_c)\cdot\mathbf{uu}) \tag{5.20}$$

They are numerical coefficients determined entirely by the structure of the initial equations and the solution of the eigenvalue problem of the linearized operator.

In addition to c, eq. (5.19) still features the so far undetermined parameter γ_1. Now, one may notice from the first equation (5.7) that the important quantity determining $\mathbf{x}$ to the dominant order is $\varepsilon\mathbf{x}_1$ rather than $\mathbf{x}_1$. We therefore introduce the normalized amplitude

$$z = \varepsilon c \tag{5.21}$$

We further multiply both sides of (5.19) by ε^2 and eliminate γ_1 and τ_1 in favor of the initial physical parameters $\lambda - \lambda_c$ and t, through inversion (again to the dominant order) of (5.8) and (5.9). Eq. (5.19) is then transformed to

$$dz/dt = (\lambda - \lambda_c)P_1 z - P_2 z^2 \tag{5.22}$$

As long as $P_2 \neq 0$ it admits two fixed points

$$\left.\begin{aligned} z_{s0} &= 0 \\ z_{s1} &= (\lambda - \lambda_c)\frac{P_1}{P_2} \end{aligned}\right\} \tag{5.23}$$

Plotting these solutions in terms of λ one finds the graphs (a) or (a') of Fig.

5.1 according to whether P_1/P_2 is positive or negative, these being the two possible forms of transcritical bifurcation. The stability of the branches is found by setting

$$z = z_s + \zeta$$

and by linearizing in ζ:

$$d\zeta/dt = [(\lambda - \lambda_c)P_1 - 2P_2 z_s]\zeta$$

yielding a characteristic exponent

$$\omega = (\lambda - \lambda_c)P_1 - 2P_2 z_s$$

or:

$$\left. \begin{array}{ll} \omega = (\lambda - \lambda_c)P_1 & \text{for } z_s = z_{s0} = 0 \\[2ex] \omega = -(\lambda - \lambda_c)P_1 & \text{for } z_s = z_{s1} = (\lambda - \lambda_c)\dfrac{P_1}{P_2} \end{array} \right\} \quad (5.24)$$

To evaluate the sign of P_1 we differentiate both sides of the eigenvalue equation (4.2a) with respect to λ and evaluate the result at the critical point $\lambda = \lambda_c$,

$$\mathscr{L}_\lambda(\lambda_c) \cdot \mathbf{u} + \mathscr{L}(\lambda_c) \cdot \mathbf{u}_{\lambda_c} = \left[\frac{d\omega(\lambda)}{d\lambda} \right]_{\lambda_c} \mathbf{u} + \omega_c \mathbf{u}_{\lambda_c}$$

where $\mathbf{u} = \mathbf{u}(\lambda_c)$ is the critical eigenvector. Taking the scalar product of both sides of this equation with $\mathbf{u}^+$ and using (5.17) we obtain, recalling that $\omega_c = 0$,

$$(\mathbf{u}^+, \mathscr{L}_\lambda(\lambda_c) \cdot \mathbf{u}) = \left[\frac{d\omega(\lambda)}{d\lambda} \right]_{\lambda_c} (\mathbf{u}^+, \mathbf{u})$$

or finally, using the first relation (5.20),

$$P_1 = \left[\frac{d\omega(\lambda)}{d\lambda} \right]_{\lambda_c} \quad (5.25)$$

This is nothing but the expression appearing in the transversality condition (5.6). We may assume without loss of generality that $(d\omega/d\lambda)_{\lambda_c} > 0$, hence $P_1 > 0$. Eqs. (5.24) imply then that the nontrivial (bifurcating) branch z_{s1} is stable in the range $\lambda > \lambda_c$, in which it is supercritical, and unstable in the range $\lambda < \lambda_c$, in which it is subcritical.

The above analysis provides one with the constructive proof of theorems 1 and 2 of Section 5.1 for transcritical bifurcation. A number of important points that have emerged from the analysis deserve special mention.

(i) Eq. (5.22) is *universal*, in the sense that its form is independent of the detailed structure of the underlying model. The latter enters only through the numerical values of the coefficients P_1 and P_2. All dynamical systems undergoing a transcritical bifurcation are thus described, to the dominant order, by this equation. For this reason we shall refer to (5.22) as the *normal form* of transcritical bifurcation.

(ii) While the original system is, in general, a multivariate system, eq. (5.22) is a single scalar equation. In the vicinity of the bifurcation there is thus a dramatic reduction in the description of a dynamical system in the sense that the only relevant variable appears to be the normalized amplitude z, which will be referred to for this reason as the *order parameter* (Landau and Lifshitz, 1959b). We call the reduced subspace of the full phase space in which the dynamics of the order parameter is taking place the *center manifold* (Guckenheimer and Holmes, 1983). All other variables follow z passively according to eq. (5.13). This enhances enormously the power and range of applicability of bifurcation analysis as well as the importance of low-dimensional dynamical systems.

5.4 The amplitude equation: pitchfork bifurcation

The analysis of the preceding section obviously fails when $P_2 = 0$. At first sight this would seem to be a very exceptional situation, but on a closer examination of eq. (5.20) one realizes that it may be typical in systems displaying symmetries. Such symmetries are manifested most naturally in the presence of spatial degrees of freedom as we see further in Chapter 6, but may well subsist when the reduction to a finite number of variables described in Section 3.1 is operated.

Be it as it may, an immediate consequence of $P_2 = 0$ in eq. (5.19) or (5.22) is that the normal form equation is reduced to its linear part which shows unphysical runaway behavior beyond bifurcation unless, of course, it reduces to a trivial identity. This will be so provided that

$$\left. \begin{array}{l} \gamma_1 = 0 \\ \partial c / \partial \tau_1 = 0 \end{array} \right\} \quad (5.26a)$$

The solvability condition (5.18) being still (trivially) satisfied, one may solve the simplified second order equation (eq. (5.15)) to obtain

$$\mathbf{x}_2 = \mathscr{L}^{-1}(\lambda_c) \cdot [- \tfrac{1}{2} c^2 \mathbf{h}_{\mathbf{xx}}(\lambda_c) \cdot \mathbf{uu}] + c_2(\tau_2) \mathbf{u} \quad (5.26b)$$

where we have added to the particular solution of the inhomogeneous eq. (5.15) the general solution of the associated homogeneous equation. To determine the amplitude c one has now to turn to the third order equation, eq. (5.12). For $\gamma_1 = 0$, no τ_1–dependence, and $\mathbf{x}_1, \mathbf{x}_2$ given by (5.13) and

(5.26b) this equation simplifies to

$$\mathscr{L}(\lambda_c) \cdot \mathbf{x}_3 = -c\gamma_2 \mathscr{L}_\lambda(\lambda_c) \cdot \mathbf{u} + \frac{\partial c}{\partial \tau_2} \mathbf{u} - \tfrac{1}{6}c^3 \mathbf{h}_{\mathbf{xxx}}(\lambda_c) \cdot \mathbf{uuu}$$
$$+ \tfrac{1}{2}c^3 \mathbf{h}_{\mathbf{xx}}(\lambda_c) \cdot \mathbf{u}\{\mathscr{L}^{-1}(\lambda_c) \cdot [\mathbf{h}_{\mathbf{xx}}(\lambda_c) \cdot \mathbf{uu}]\}$$
$$- cc_2 \mathbf{h}_{\mathbf{xx}}(\lambda_c) \cdot \mathbf{uu} = \mathbf{q}_3(\mathbf{u}, c, \lambda_c) \tag{5.27}$$

These relations constitute again an inhomogeneous set of equations, this time for $\mathbf{x}_3$. To compute this unknown vector one must ensure the invertibility of $\mathscr{L}(\lambda_c)$ through the solvability condition (cf. eq. (5.18))

$$(\mathbf{u}^+, \mathbf{q}_3) = 0 \tag{5.28}$$

Substituting $\mathbf{q}_3$ from (5.27) and following the lines of the preceding section one finally ends up with

$$\partial c / \partial \tau_2 = \gamma_2 P_1 c - P_3 c^3 \tag{5.29}$$

where P_1 is given by the first relation (5.20) and

$$P_3 = \frac{-1}{(\mathbf{u}^+, \mathbf{u})} (\mathbf{u}^+, [\tfrac{1}{6}\mathbf{h}_{\mathbf{xxx}}(\lambda_c) \cdot \mathbf{uuu} - \tfrac{1}{2}\mathbf{h}_{\mathbf{xx}}(\lambda_c) \cdot \mathbf{u}\{\mathscr{L}^{-1}(\lambda_c) \cdot [\mathbf{h}_{\mathbf{xx}}(\lambda_c) \cdot \mathbf{uu}]\}])$$
$$\tag{5.30}$$

Notice that in view of $P_2 = 0$, the solution of the homogeneous equation does not contribute to the solvability condition. Switching to the normalized amplitude $z = \varepsilon c$ and eliminating γ_2 and τ_2 in favor of the physical parameters $\lambda - \lambda_c$ and t through the inversion of eqs. (5.8) and (5.9) (now with $\gamma_1 = 0$ and no τ_1–dependence) one can write (5.29) in the more suggestive form

$$dz/dt = (\lambda - \lambda_c)P_1 z - P_3 z^3 \tag{5.31}$$

As long as $P_3 \neq 0$ this equation may admit up to three fixed points,

$$z_{s0} = 0$$
$$z_{s,\pm} = \pm [(\lambda - \lambda_c)P_1/P_3]^{1/2} \tag{5.32}$$

Plotting these solutions against λ one finds the graphs (b) or (b') of Fig. 5.1, according to whether P_1/P_3 is positive or negative, these being the two possible forms of pitchfork bifurcation. The stability of the branches can be studied exactly as in the previous section, the result being that supercritical branches are stable and subcritical ones unstable. We have thus extended the constructive proof of theorems 1 and 2 of Section 5.1 to include the case of pitchfork bifurcation.

The comments at the end of Section 5.3 apply fully to eq. (5.31) as well: it is a universal equation, effectively reducing the initial multivariate problem to a single variable. All dynamical systems operating in the vicinity of a pitchfork bifurcation can be cast into eq. (5.31), which for this reason can be regarded as the *normal form* for this type of bifurcation.

A new feature not encountered in the transcritical bifurcation is that at the stationary state the order parameter z depends on the distance from bifurcation in a nonanalytic manner (eq. (5.32)). This is the mathematical manifestation of the qualitative change of behavior across a pitchfork bifurcation and, more specifically, of the fact that for λ as close to λ_c as desired one finds two branches of coexisting states having identical stability properties. This is reminiscent of many of the experimental data surveyed in Chapter 1.

5.5 Limit point bifurcation

The basis of our analysis of the preceding sections was the assumption, spelled out explicitly in the statement of theorem 1 of Section 5.1, that the reference state remains an exact solution of the system equations in a neighborhood of the bifurcation point. This means, in particular, that $\mathbf{F}(\mathbf{X}_s, \lambda)$ and all its derivatives with respect to λ vanish at $\lambda = \lambda_c$. In this section we examine the type of behavior that can take place when these conditions are not satisfied. To simplify notation we limit ourselves to one variable.

Let X_s be a reference state that exists, say, up to the value λ_c of the control parameter λ, $F(X_{sc}, \lambda_c) = 0$. Suppose furthermore that at λ_c it loses its stability through a simple real eigenvalue that becomes zero, $(\partial F/\partial X)_c = 0$, where the subscript c indicates that the derivatives are to be evaluated at X_{sc} and λ_c. We expand the evolution equation

$$dX/dt = F(X, \lambda) \tag{5.33}$$

around (X_{sc}, λ_c)

$$\frac{dX}{dt} = F(X_{sc}, \lambda_c) + \left(\frac{\partial F}{\partial X}\right)_{X_{sc}, \lambda_c} (X - X_{sc}) + \left(\frac{\partial F}{\partial \lambda}\right)_{X_{sc}, \lambda_c} (\lambda - \lambda_c)$$
$$+ \tfrac{1}{2}\left(\frac{\partial^2 F}{\partial X^2}\right)_{X_{sc}, \lambda_c} (X - X_{sc})^2 + \cdots \tag{5.34}$$

The first two terms vanish on the grounds of our assumptions. Introducing the notation

$$\left. \begin{aligned} z &= X - X_{sc} \\ \mu &= (\partial F/\partial \lambda)_{X_{sc}, \lambda_c}(\lambda - \lambda_c) \\ q_2 &= -\tfrac{1}{2}(\partial^2 F/\partial X^2)_{X_{sc}, \lambda_c} \end{aligned} \right\} \tag{5.35}$$

we may transform, to the dominant order, the original equation into

$$dz/dt = \mu - q_2 z^2 \tag{5.36}$$

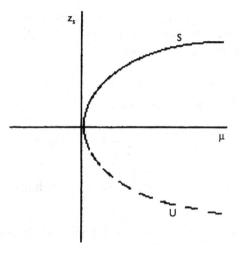

Fig. 5.2 Dependence of the amplitude z_s of the solution on the parameter μ in the vicinity of a limit point bifurcation.

This equation admits a pair of fixed points,

$$z_{s,\pm} = \pm\,(\mu/q_2)^{1/2} \tag{5.37}$$

which exist in the region $\mu > 0$ if q_2 is positive and $\mu < 0$ if q_2 is negative. They are plotted against μ (for $q_2 > 0$) in Fig. 5.2. To check stability we set $z = z_{s,\pm} + \zeta$ and linearize (5.36) with respect to ζ, getting

$$\mathrm{d}\zeta/\mathrm{d}t = -2q_2 z_{s,\pm}\zeta$$

This shows that if $q_2 > 0$, the positive branch $z_{s,+}$ is stable and the negative one $z_{s,-}$ is unstable. We conclude that as μ decreases from positive values, the stable and unstable branches 'collide' at $\mu = 0$ and subsequently are annihilated. For this reason we call $\mu = 0$ a *limit point*, or fold. Its presence signals once again the appearance of singularities, as illustrated by the nonanalytic dependence of $z_{s,\pm}$ on μ in eq. (5.37).

The above analysis extends straightforwardly to the multivariate case. One still ends up with an equation of the form (5.36) satisfied by a suitable combination of the original variables, which can therefore be regarded as the normal form of a dynamical system operating in the vicinity of a limit point bifurcation (Guckenheimer and Holmes, 1983). A widely encountered realization of this type of bifurcation is when two fixed points that coalesce at $\mu = 0$ behave for $\mu \neq 0$ as a stable node and as a saddle, in the sense of the classification of Sections 4.3 and 4.5. For this reason it is also referred to in the literature as a *saddle-node bifurcation*.

5.6 Kinetic potential, sensitivity, structural stability

The reduction of bifurcation of steady-state solutions at a simple real

eigenvalue to a one-dimensional dynamics entails the interesting consequence that the normal form equations (5.22), (5.31) or (5.36) derive
necessarily from a potential,

$$dz/dt = -\partial U/\partial z \qquad (5.38)$$

with, up to an arbitrary constant,

$$U = -(\lambda - \lambda_c)P_1\frac{z^2}{2} + P_2\frac{z^3}{3} \quad \text{(transcritical bifurcation)} \quad (5.39a)$$

$$U = -(\lambda - \lambda_c)P_1\frac{z^2}{2} + P_3\frac{z^4}{4} \quad \text{(pitchfork bifurcation)} \qquad (5.39b)$$

$$U = -\mu z + q_2\frac{z^3}{3} \quad \text{(limit point bifurcation)} \qquad (5.39c)$$

The structure of the potential U bears a straightforward relation with the
fixed points of the dynamical system and their stability. Indeed, by (5.38)
the fixed points are extrema of U and vice versa

$$(\partial U/\partial z)_{z_s} = 0 \qquad (5.40)$$

Setting $z = z_s + \zeta$ and linearizing (5.38) with respect to ζ one may also
cast the stability problem in the form

$$d\zeta/dt = -(\partial^2 U/\partial z^2)_{z_s}\zeta$$

It follows that:

if z_s is a minimum of U, $(\partial^2 U/\partial z^2)_{z_s} > 0$ and z_s is stable;
if z_s is a maximum of U, $(\partial^2 U/\partial z^2)_{z_s} < 0$ and z_s is unstable.

Fig. 5.3 describes on the basis of eq. (5.39b), the structural changes of U
as the underlying dynamical system undergoes a pitchfork bifurcation,
choosing, to fix ideas, $P_1 > 0$ and $P_3 > 0$. We notice a striking analogy
with the Landau theory of order–disorder transitions or the van der
Waals theory of liquid–vapor transition (Landau and Lifshitz, 1959b;
Stanley, 1971; Ma, 1976) with the notable difference that U is here not a
thermodynamic potential but, rather, a *kinetic* one determined by the
dynamics. The analogy also extends to the behavior of z versus λ. Indeed,
the nonanalytic dependence in eq. (5.32)

$$z_s \sim \pm|\lambda - \lambda_c|^{1/2} \qquad (5.41)$$

is also encountered in Landau theory, where the role of the control
parameter is played by the temperature. In the theory of critical

phenomena the exponent 1/2, referred to as *critical exponent*, is character-istic of a *mean field description*, which is now known to break down in the immediate vicinity of the critical point (Ma, 1976). In dynamical systems with a finite number of degrees of freedom the mean field exponent is exact, but qualitative changes may be expected in spatially extended systems as discussed further in Chapter 6.

Despite the above mentioned appealing analogies one should refrain from identifying bifurcations in nonlinear systems under constraint with equilibrium phase transitions. The microscopic basis of these two classes of phenomena is indeed very different. When fluctuations are incorpor-ated in the description these differences show up in a number of properties such as, for instance, the parameter values at which the coexisting attractors z_- and z_+ are equiprobable (Nicolis and Turner, 1977; Nicolis and Lefever, 1977). Another major difference is that, contrary to equilibrium phase transitions which are mediated entirely by the inter-molecular forces, bifurcations in systems under constraint bring about space and time scales which are macroscopic.

An alternative, interesting vision of the three fundamental bifurcations at a simple real eigenvalue can be achieved when one realizes that the limit point bifurcation can actually be viewed as the result of an *imperfection* perturbing a transcritical or a pitchfork bifurcation. Consider, for instance, the equation

$$dx/dt = ax^3 + bx^2 + cx + d$$

which reduces to the normal form of pitchfork bifurcation for $b = d = 0$

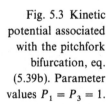
Fig. 5.3 Kinetic potential associated with the pitchfork bifurcation, eq. (5.39b). Parameter values $P_1 = P_3 = 1$.

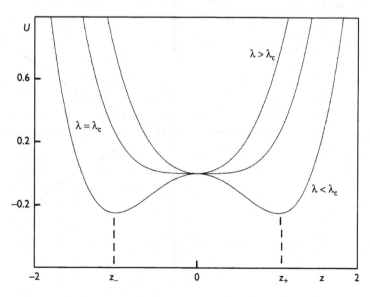

and to the limit point bifurcation for $a = c = 0$. What happens when all these terms coexist? We notice that the coefficient a can be eliminated, as long as it is not strictly zero, by a change of time scale and the quadratic term can likewise be eliminated by a shift of the variable x. One then obtains the canonical form of the cubic equation

$$\mathrm{d}z/\mathrm{d}\tau = -z^3 + \lambda z + \mu \qquad (5.42)$$

where $\tau = |a^{-1}|t, z = x - b/3|a|$, λ and μ are combinations of the original parameters and the negative sign in front of z^3 accounts for stability.

We know from elementary algebra that eq. (5.42) can have up to three fixed points. Moreover, as the parameters vary the three solutions merge and we are left with only one (real) solution. One can determine a relation between parameters separating these two regimes, specifically,

$$-4\lambda^3 + 27\mu^2 = 0 \qquad (5.43)$$

These curves are represented in parameter space in Fig. 5.4. The region of three real solutions ends at a point (the origin in the figure), in which there is a singular dependence of λ on μ. This is known as a *cusp singularity* (Thom, 1962).

Figs. 5.5(a) and (b) provide two different views of the dependence of the solutions on the parameters. In Fig. 5.5(a) z_s is plotted against μ for fixed λ. The resulting S-shaped curve indicates the coexistence of multiple

Fig. 5.4 The regions of existence of one and three real solutions of eq. (5.42) in parameter space.

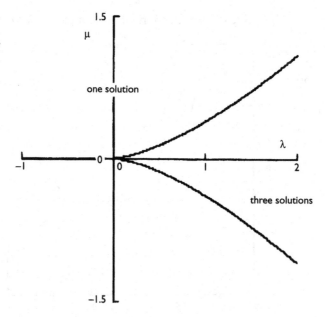

solutions for a certain range of parameter values. Stability analysis along the by now familiar lines shows that two of these branches are simultaneously stable. The bistability region ends at the limit points μ_1 and μ_2 in the vicinity of which we see the behavior shown in Fig. 5.2. Under these conditions an increase of μ beyond μ_1 and up to μ_2, followed by a variation in the opposite direction, will lead to a *hysteresis cycle* reminiscent of some of the experimental facts surveyed in Chapter 1.

In Fig. 5.5(b) z_s is plotted against λ at fixed μ. We now obtain two disjoint curves: one, (1), defined for all values of λ, and the other, (2), defined for $\lambda \geq \bar{\lambda}$ and exhibiting a limit point singularity at $\bar{\lambda}$. For $\lambda < \bar{\lambda}$ only one stable solution is available, but for $\lambda > \bar{\lambda}$ we have bistability as before. We realize that for no nonvanishing $|\mu|$, however small, can the pitchfork bifurcation (Fig. 5.1(b) and Section 5.4) be observed. The *imperfection* expressed by the presence of μ therefore destroys this bifurcation. In contrast the limit point bifurcation proves to be robust, since it is recovered in both Figs. 5.5(a) and (b). But if λ and μ are varied simultaneously there will always be a particular combination of values ($\mu = 0$, λ going through zero in our case) for which the pitchfork bifurcation will be recovered, as the system will be able to traverse the cusp singularity in a symmetric fashion.

The above discussion illustrates the deep concept of *structural stability*, to which we have already alluded in the preceding chapters. It shows that certain phenomena, such as the pitchfork bifurcation, occur only if the parameters present satisfy at least one equality. Inasmuch as in a physical system such a strict requirement will be difficult to meet, we expect that these phenomena will disappear under slight changes of parameter values: we describe them as being *structurally unstable*. On the other hand there exist other phenomena, like the limit point bifurcation, that persist (even though they may be shifted) under changes of the control parameters affecting the structure of the evolution laws: these we call *structurally stable*. To account for the full set of potentialities of a given dynamical

Fig. 5.5 Effect of parameters in the bifurcation of steady state solutions of eq. (5.42). (a) Hysteretic behavior of the solution at fixed λ (here $\lambda = 1$), as the parameter μ is varied. The limit point bifurcation remains robust. (b) Destruction of pitchfork bifurcation when the parameter μ, acting as an imperfection, is not identically zero (here $\mu = 0.05$).

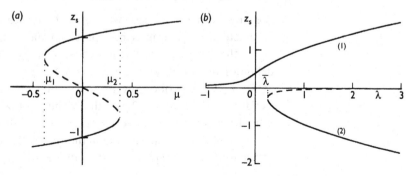

system, including the structurally unstable situations delimiting different types of structurally stable behavior, it is essential to be able to control a sufficient number of parameters. This number depends on the type of phenomenon to be described and on the degree of nonlinearity – for instance, it is equal to two for the full classification of bifurcation phenomena in the presence of a cubic nonlinearity (eq. (5.42)). A more comprehensive study of this question brings us to the concept of *universal unfolding* and to *catastrophe theory* (Thom, 1962; Dumortier, 1991) whose technical details are beyond the scope of the present monograph.

In many physical systems imperfections of the type studied above arise in the presence of external fields. In chemical reactions, when certain types of external constraint related to pumping of material in the reactor are present, imperfections may likewise arise. This is nicely illustrated by the second Schlögl model (eq. (2.45b))

$$\mathrm{d}X/\mathrm{d}t = k_1 A X^2 - k_2 X^3 + k_4 B - k_3 X \qquad (5.44)$$

Introducing the new quantities

$$z = \frac{X}{(k_1 A/3k_2)} - 1, \qquad \tau = k_2 (k_1 A/3k_2)^2 t$$

$$\lambda = 3 - \frac{k_3}{k_2} \frac{1}{(k_1 A/3k_2)^2}, \qquad \mu = \frac{-k_3}{k_2} \frac{1}{(k_1 A/3k_2)^2} + \frac{k_4 B}{k_2 (k_1 A/3k_2)^3} + 2$$

$$(5.45)$$

one can check straightforwardly that the initial four-parameter system reduces to a two-parameter one, described by eq. (5.42). The 'imperfection' μ is, in principle, present as a result of pumping A and B in the system. But there exists a relation between A and B, given by $\mu = 0$ in eq. (5.45), for which its effect on the evolution of the concentration variable is canceled. The system then undergoes a pitchfork bifurcation.

5.7 The Hopf bifurcation

We now turn to the case where, at the critical value of the control parameter $\lambda = \lambda_c$, the imaginary part Ω_c of the (simple) critical eigenvalue ω_c of the linearized operator (whose real part vanishes by the very definition of criticality) is nonzero. According to eq. (5.3) this entails that at λ_c, **x** varies periodically on a fast time scale, given by $2\pi/\Omega_c$. As explained in Section 5.2, in the vicinity of λ_c an additional slow time-dependence is expected (critical slowing down). To account for all these possibilities we have to adapt the multiscale perturbation expansion of Section 5.2 and replace eq. (5.9) by the new scaling relation

$$\frac{\mathrm{d}}{\mathrm{d}t} = \Omega_c \frac{\partial}{\partial T} + \varepsilon \frac{\partial}{\partial \tau_1} + \varepsilon^2 \frac{\partial}{\partial \tau_2} + \cdots \qquad (5.46)$$

fast scale

With eq. (5.46), eqs. (5.1) may now be solved perturbatively, by expanding as before $\mathbf{x}$ and λ in power series of ε (eqs. (5.7), (5.8)). We briefly outline the structure of the equations of the first few significant orders.

A $O(\varepsilon)$

We obtain an additional contribution to (5.10) arising from the presence of the fast time scale:

$$[\Omega_c \frac{\partial}{\partial T}\mathbf{1} - \mathscr{L}(\lambda_c)] \cdot \mathbf{x}_1 = 0 \qquad (5.47)$$

This homogeneous system of equations (identical to the problem of linear stability analysis) admits solutions of the form

$$\mathbf{x}_1 = c(\tau_1, \tau_2, \ldots)\mathbf{u}\,e^{iT} + cc \qquad (5.48)$$

where $\mathbf{u}$ is given by eq. (5.4) and is, therefore, the null eigenvector of the operator $\mathbf{J}_c$ defined in eq. (5.5). The amplitude c remains undetermined at this stage and is allowed to depend on the slow time scales, which do not enter explicitly in eq. (5.47).

B Higher orders

To $O(\varepsilon^2)$, $O(\varepsilon^3)$ etc., substitution of (5.7), (5.8) and (5.46) into eqs. (5.1) will give rise to a sequence of linear *inhomogeneous* equations of the form

$$\mathbf{J}_T \cdot \mathbf{x}_j = \left[\Omega_c \frac{\partial}{\partial T}\mathbf{1} - \mathscr{L}(\lambda_c)\right] \cdot \mathbf{x}_j = \mathbf{q}_j(\mathbf{x}_{j-1}, \ldots, \mathbf{x}_1) \quad j \geqslant 2 \quad (5.49)$$

In order to determine the solution $\mathbf{x}_j$ one needs to ensure that the operator $\mathbf{J}_T$ is invertible. This is, however, not necessarily true since according to eqs. (5.47) this operator has a nontrivial null space. To ensure invertibility a solvability condition must be fulfilled, in the form of orthogonality of $\mathbf{q}_j$ and the null eigenvector of the adjoint operator $\mathbf{J}_T^+$. Orthogonality and adjointness must here be defined in an extended space which, in addition to being the finite vector space considered in the preceding sections, includes 2π-periodic functions in time. In this functional space (Friedman, 1956) the scalar product will be defined as

$$(\mathbf{a}(T), \mathbf{b}(T)) = \int_0^{2\pi} dT \mathbf{a}^*(T) \cdot \mathbf{b}(T) \qquad (5.50)$$

where $\mathbf{a}^* \cdot \mathbf{b}$ stands for the classical scalar product, $\mathbf{a}^*$ being the complex conjugate of $\mathbf{a}$. The adjoint operator $\mathbf{J}_T^+$ can be identified by the property

$$(\mathbf{a}(T), \mathbf{J}_T \cdot \mathbf{b}(T)) = (\mathbf{J}_T^+ \cdot \mathbf{a}(T), \mathbf{b}(T))$$

or, performing integration by parts in time,

$$\mathbf{J}_T^+ = -\Omega_c \frac{\partial}{\partial T} \mathbf{1} - \mathscr{L}^+(\lambda_c) \qquad (5.51)$$

where $\mathscr{L}^+(\lambda_c)$ has the same significance as in the preceding sections. The null eigenvector of this operator is of the form

$$\mathbf{u}^+ e^{iT},$$

$\mathbf{u}^+$ being the null eigenvector of the adjoint of $\mathbf{J}_c$ in eq. (5.5).

Applied to the second order equation the solvability condition yields

$$\frac{\partial c}{\partial \tau_1} = 0, \gamma_1 = 0 \qquad (5.52)$$

owing to the fact that $(\mathbf{u}^+ e^{iT}, \mathbf{h}_{\mathbf{xx}} \cdot \mathbf{u} e^{iT} \mathbf{u} e^{iT})$ vanishes identically. The second order equation can then be solved, the result being a π-periodic solution (second harmonic of (5.48)) plus a time-independent part. Since the second order solvability condition is trivial this solution still features the amplitude c. To determine it one must therefore proceed to the third order in ε. The solvability condition to this order now gives a nontrivial result which, on transforming back to the original variables and parameters, reads

$$dz/dt = (\lambda - \lambda_c)P_1 z - P_3 |z|^2 z \qquad (5.53)$$

where the coefficients P_1 (still given by eq. (5.25)) and P_3 are in general complex quantities $P_1 = P_1' + iP_1'', P_3 = P_3' + iP_3''$. Eq. (5.53) constitutes the normal form of the Hopf bifurcation (Marsden and McCracken, 1976; Guckenheimer and Holmes, 1983).

To solve eq. (5.53) it is convenient to introduce polar coordinates

$$z = r e^{i\phi} \qquad (5.54)$$

Separating real and imaginary parts one obtains

$$\left.\begin{array}{l} dr/dt = (\lambda - \lambda_c)P_1'r - P_3'r^3 \\[12pt] d\phi/dt = (\lambda - \lambda_c)P_1'' - P_3''r^2 \end{array}\right\} \qquad (5.55)$$

The first relation is independent of the phase variable ϕ. It can be handled as eq. (5.31), characteristic of the pitchfork bifurcation, except that r must be always nonnegative. Taking up the results of Section 5.4 we conclude that at $\lambda = \lambda_c$ r undergoes bifurcation from the trivial solution $r_0 = 0$ to the nontrivial one

$$r_s = \left[\frac{(\lambda - \lambda_c)P_1'}{P_3'}\right]^{1/2} \qquad (5.56)$$

which extends in the domain $\lambda > \lambda_c$ if $P_1'/P_3' > 0$ and in the domain $\lambda < \lambda_c$ if $P_1'/P_3' < 0$. Furthermore, supercritical branches can be proven to be stable and subcritical ones unstable.

At the level of the first eq. (5.55) the solutions just found are time-independent, but at the level of the z-variable (eq. (5.54)) they are time-periodic since from the second eq. (5.55) one has in the long time limit,

$$d\phi/dt = (\lambda - \lambda_c)P_1'' - P_3''r_s^2$$

or

$$\phi = \phi_0 + \left(P_1'' - \frac{P_3''P_1'}{P_3'}\right)(\lambda - \lambda_c)t = \phi_0 + \Delta\Omega t \qquad (5.57)$$

The frequency $\Delta\Omega$ of the z-oscillation corrects the value Ω_c (eq. (5.49)) by terms of the order of $\lambda - \lambda_c$. In the phase space (Im z, Re z) the trajectories are winding toward the invariant curve $|z| = r_s$, which we have referred to as the *limit cycle* (Fig. 5.6). The representative point moves on this attractor at an angular velocity equal to $\Delta\Omega$. We have thus produced a constructive proof of the bifurcation of time-periodic solutions in the vicinity of a criticality corresponding to a simple pair of complex eigenvalues. Notice that the period and the amplitude of the oscillations are *intrinsic* in the sense that they are determined entirely by the evolution laws, independent of the initial conditions. Many of the experimental data surveyed in Chapter 1, such as chemical oscillations and biological rhythms, feature this property. This is to be contrasted with oscillations in conservative systems, whose characteristics depend on the initial conditions.

Just like in Sections 5.3–5.5 one may remark that eqs. (5.53) or (5.55) are *universal*: they constitute the *normal form* of any dynamical system operating in the vicinity of a Hopf bifurcation. One notable difference

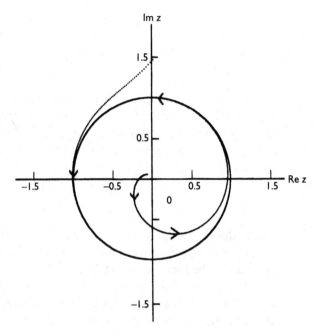

Fig. 5.6 Phase portrait
beyond the Hopf
bifurcation, as
deduced from the
normal form equation
(5.53). Parameter
values: $P_1 = P_3 = 1$,
$\lambda - \lambda_c = 1$, frequency
at criticality $\Omega_c = 1$.

with the bifurcation at a real eigenvalue is, however, that the order parameter z is complex. The subspace of phase space on which the reduced dynamics is taking place is, therefore, two-dimensional. An explicit calculation of the periodic solutions of the Brusselator model following the above scheme is performed in Appendix A2.

5.8 Cascading bifurcations

Bifurcations at simple eigenvalues analyzed in the preceding sections are well understood since they can be reduced to a *local* problem around the reference state $\mathbf{X}_s$, assumed to be known explicitly. Furthermore, since they are fully described by a scalar or a two-component vector order parameter, the phase portraits which they generate can be classified completely (Andronov *et al.*, 1966) and can only give rise to structurally stable attractors in the form of fixed points or limit cycles. As shown in Chapter 3 this is mainly due to the topological constraints imposed on the trajectories embedded in a one- or two-dimensional phase space, and more particularly to the fact that self-intersections are not allowed.

What happens beyond the first (local) bifurcation from $\mathbf{X}_s$ is, on the other hand, generally a global problem. For instance, the first bifurcating solution emerging at λ_{c1} may lose its stability at some critical value λ'_{c1} where a new solution branch may bifurcate. But typically λ'_{c1} will be at finite distance from λ_{c1}. The problem of extending the *primary* bifurcating branch (known explicitly only near λ_{c1}) up to λ'_{c1} and of constructing the new *secondary* bifurcating branch becomes, then, a highly nonlinear and generally intractable problem.

One way out of this difficulty is to identify in the system one or several control parameters in addition to λ and to vary them until critical situations of a new kind are reached. Fig. 5.7 illustrates the idea for two control parameters λ and μ. At $\mu = \mu_1$ we obtain an ordinary one-parameter bifurcation diagram displaying a number of primary branches a_1, b_1 ... as well as a secondary branch a_2. Both λ_{c1}, λ_{c2} and the amplitudes of the solutions depend now on μ. One may therefore imagine that by selecting μ in an appropriate manner, for some value $\mu = \mu_2$ the distance between λ_{c1}, λ_{c2} and λ'_{c1} will become smaller and the branches a_1, b_1, a_2 will begin to interact. On further varying μ one might even reach a situation corresponding to parameter values $(\bar{\lambda}, \bar{\mu})$ in which the two primary bifurcation points will coalesce. It may be expected that at this point many different branches of bifurcating solutions will emerge.

Fig. 5.7 Illustration of the mechanisms by which a global problem related to a secondary bifurcation (a), reduces to a local one (b), in the vicinity of a high codimension bifurcation (c), thanks to the simultaneous control of two parameters λ and μ.

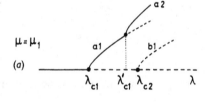

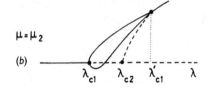

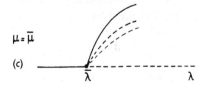

However, on slightly varying λ and μ from this situation the branches will split and a secondary bifurcation will take place (Bauer, Keller and Reiss, 1975; Keener, 1976). The point is that since $\mathbf{X}_s$ and $(\bar{\lambda}, \bar{\mu})$ are known explicitly from algebra and linear stability analysis, the initial global problem has been reduced, in principle, to a local one. We refer to it as *higher codimension* (here two) bifurcation, in the sense that the parameters must satisfy more than one strict equality. Physically speaking such bifurcations should arise either in the presence of symmetries, or when two or more mechanisms of instability interfere simultaneously with the dynamics.

The simplest cases of higher codimension bifurcations are:

(i) *Two real eigenvalues go through zero*
At criticality one gets a double zero eigenvalue $\omega_0 = 0$ of $\mathscr{L}$, that is to say an eigenvalue of algebraic multiplicity $\mu_0 = 2$. In the terminology of Section 4.1 two subcases may arise, depending on the value of the geometric multiplicity ν_0.

(i.1) *Geometric multiplicity $\nu_0 = 2$*
The transformed linearized matrix $\mathscr{L}$ (eq. (4.11a)) will contain at criticality two Jordan blocks of order $n_0 = 1$ containing a zero matrix element (the vanishing eigenvalue), which can be grouped into the 2×2 Jordan block

$$\mathbf{J}_c = \begin{pmatrix} 0 & 0 \\ 0 & 0 \end{pmatrix} \qquad (5.58a)$$

Obviously, $\binom{1}{0}$ and $\binom{0}{1}$ are two linearly independent null eigenvectors of $\mathbf{J}_c$.

(i.2) *Geometric multiplicity $\nu_0 = 1$*
The transformed linearized matrix will contain at criticality a Jordan block in the form (cf. eq. (4.11b)):

$$\mathbf{J}_c = \begin{pmatrix} 0 & 1 \\ 0 & 0 \end{pmatrix} \qquad (5.58b)$$

The only null eigenvector is now $\mathbf{u}_1 = \binom{1}{0}$. To complete the basis of the 2×2 critical subspace one introduces generalized eigenvectors (Sattinger, 1972; Iooss and Joseph, 1980)

$$\mathbf{J}_c \cdot \mathbf{u}_2 = \mathbf{u}_1, \cdots \qquad (5.59)$$

In the present case there is only one such additional eigenvector. Notice that by (5.59), $\mathbf{J}_c^2 \cdot \mathbf{u}_2 = 0$.

(ii) *One real eigenvalue goes through zero and one pair of complex conjugate eigenvalues (with Im $\omega \neq 0$) has a vanishing real part*

At criticality the eigenvalues of $\mathscr{L}$ remain distinct. The corresponding Jordan block will be (cf. eq. (4.11c)):

$$\mathbf{J}_c = \begin{pmatrix} 0 & 0 & 0 \\ 0 & 0 & \omega \\ 0 & -\omega & 0 \end{pmatrix} \qquad (5.60)$$

(iii) *Three real eigenvalues go through zero*

At criticality one gets a triple zero eigenvalue of $\mathscr{L}$. As in (i) one may distinguish various cases, depending on the value of the geometric multiplicity ν_0. For $\nu_0 = 1$ the transformed linearized matrix contains a Jordan block of the form (cf. eq. (4.11b)):

$$\mathbf{J}_c = \begin{pmatrix} 0 & 1 & 0 \\ 0 & 0 & 1 \\ 0 & 0 & 0 \end{pmatrix} \qquad (5.61)$$

(iv) *Two pairs of complex conjugate eigenvalues (with Im $\omega \neq 0$) have vanishing real parts*

As in case (ii) the eigenvalues of $\mathscr{L}$ remain distinct at criticality. The corresponding Jordan block in the transformed linearized matrix will be (cf. eq. (4.11b)):

$$\mathbf{J}_c = \begin{pmatrix} 0 & \omega_1 & & \\ -\omega_1 & 0 & & 0 \\ & & 0 & \omega_2 \\ & 0 & -\omega_2 & 0 \end{pmatrix} \qquad (5.62)$$

As in the preceding sections of this chapter, the perturbative approach can again be applied to derive the equations to which the dynamics is reduced. Let us illustrate the procedure for the criticality of type (i.1). We want again to solve eqs. (5.1), this time near $(\bar{\lambda}, \bar{\mu})$. Since we dispose of two control parameters we enlarge the expansion (5.7)–(5.9) to

$$\left. \begin{aligned} \mathbf{x} &= \varepsilon \mathbf{x}_1 + \cdots \\ \lambda - \bar{\lambda} &= \varepsilon \lambda_1 + \cdots \\ \mu - \bar{\mu} &= \varepsilon \mu_1 + \cdots \\ d/dt &= \varepsilon\, \partial/\partial \tau_1 + \cdots \end{aligned} \right\} \qquad (5.63)$$

Substituting into (5.1) we obtain, to the first order in ε, a homogeneous equation similar to (5.10),

$$\mathscr{L}(\bar{\lambda}, \bar{\mu}) \cdot \mathbf{x}_1 = 0 \qquad (5.64)$$

The new point is, however, that owing to the type of degeneracy considered ($\mu_0 = v_0 = 2$), $\mathscr{L}(\bar{\lambda}, \bar{\mu})$ has now a *two-dimensional* null space ($\mathbf{u}_1$, $\mathbf{u}_2$). Hence eq. (5.64) admits solutions of the form

$$\mathbf{x}_1 = c_1(\tau_1, \ldots)\mathbf{u}_1 + c_2(\tau_1, \ldots)\mathbf{u}_2 \qquad (5.65)$$

To $O(\varepsilon^2)$ the analysis proceeds as in Section 5.2, except that owing to the presence of the second parameter one gets the additional term $\mathscr{L}_\mu(\bar{\lambda}, \bar{\mu}) \cdot \mathbf{x}_1$ in the right hand side of eq. (5.11). As before this relation constitutes an inhomogeneous set of equations for the components of the vector $\mathbf{x}_2$, whose solution requires a solvability condition. But since the null space of $\mathscr{L}(\bar{\lambda}, \bar{\mu})$ and of its adjoint $\mathscr{L}^+$ is now *two*-dimensional there will be two such relations,

$$(\mathbf{u}_1^+, \mathbf{q}_2) = 0, (\mathbf{u}_2^+, \mathbf{q}_2) = 0 \qquad (5.66)$$

which will provide one with a set of closed equations for the undetermined amplitudes c_1 and c_2. Introducing normalized amplitudes $z_1 = \varepsilon c_1$, $z_2 = \varepsilon c_2$, and switching back to the original parameters one finally ends up with the following augmented set of equations replacing (5.22),

$$dz_1/dt = [(\lambda - \bar{\lambda})P_1 + (\mu - \bar{\mu})P_1']z_1 - P_{11}z_1^2 - P_{12}z_1z_2 - P_{22}z_2^2 \qquad (5.67a)$$

$$dz_2/dt = [(\lambda - \bar{\lambda})Q_1 + (\mu - \bar{\mu})Q_1']z_2 - Q_{11}z_1^2 - Q_{12}z_1z_2 - Q_{22}z_2^2 \qquad (5.67b)$$

provided that the coefficients P_{ij}, Q_{ij} ($i,j = 1,2$) do not all vanish identically.

Let us analyze the bifurcation structure of the fixed points of these equations under the simplifying conditions $P_{12} = Q_{11} = Q_{22} = 0$. We notice that these conditions imply that eqs. (5.67) are invariant under the transformation $(z_1, z_2) \rightarrow (z_1, -z_2)$. Such situations may arise frequently in spatially distributed systems, as we see in the next chapter.

The fixed point solutions are:

the trivial solution

$$z_{10} = z_{20} = 0 \qquad (5.68a)$$

which we take as the reference solution
the semi-trivial solution

$$z_{1s} = \frac{(\lambda - \bar{\lambda})P_1 + (\mu - \bar{\mu})P_1'}{P_{11}}, \quad z_{2s} = 0 \qquad (5.68b)$$

the nontrivial solution

$$z'_{1s} = \frac{(\lambda - \bar{\lambda})Q_1 + (\mu - \bar{\mu})Q'_1}{Q_{12}}$$

$$z'_{2s} = \pm \left\{ \frac{z'_{1s}[(\lambda - \bar{\lambda})P_1 + (\mu - \bar{\mu})P'_1 - P_{11}z'_{1s}]}{P_{22}} \right\}^{1/2} \qquad (5.68c)$$

We notice that z_{1s} can become equal to z'_{1s} for some $\lambda^* = \lambda(\mu)$, in which case z'_{2s} vanishes according to the first relation, (5.68b) thus becoming equal to z_{2s}. At this point the interaction between the two primary branches (5.68b) and (5.68c) produces, therefore, a secondary bifurcation as illustrated in Fig. 5.8. Other secondary bifurcation phenomena, including Hopf bifurcations can arise in eqs. (5.67) when (z'_{1s}, z'_{2s}) is used as reference state.

In the example of a doubly-degenerate real eigenvalue of geometric multiplicity two considered here one arrives at a set of two normal form equations displaying two order parameters. As noticed earlier in a different context, being associated with a two-dimensional dynamical system, the phase portraits of this set of equations can be classified exhaustively. In particular, they can only admit structurally stable attractors in the form of fixed points and limit cycles.

The situation becomes more complicated in the bifurcations (ii)–(iv) listed above, where one obtains three or four coupled equations for an equal number of order parameters. Here the normal forms at which one arrives by perturbation theory can no longer be guaranteed to be universal in the vicinity of $(\bar{\lambda}, \bar{\mu})$. In particular, the effect of higher order terms or of additional parameters has not yet been fully assessed: the complete stable unfolding of the problem remains an open question. Fortunately in many situations of physical interest the type of system and the nature of the dynamics impose a particular type of unfolding. The

Fig. 5.8 Secondary bifurcation branches arising from the interaction of two primary branches near a codimension two bifurcation, eqs. (5.68).

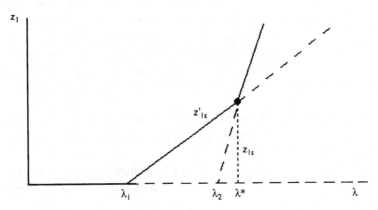

question of universality, although very important from the standpoint of mathematical completeness, thus becomes less relevant for the physics of the underlying system.

A second characteristic feature of codimension two (and, of course, also higher than two) bifurcations is that, in addition to fixed points and limit cycles, new invariant sets of the flow are generated and give rise to *global* bifurcation phenomena in the normal form, even though the latter has been established by a local theory. In the presence of at least three order parameters these global bifurcations may lead to chaotic dynamics. We may thus conclude this section by stressing that codimension two bifurcations already mark the *limits of universality* in the description of nonlinear dynamical systems. As we shall see in the next two chapters, universal laws may still emerge beyond this range. These laws are, however, of a new type as they refer to a completely different level of description.

5.9 Normal forms and resonances

An interesting, alternative approach to normal forms, which historically was actually the first one developed in the literature, is to use the Poincaré–Hartman–Grobman ideas discussed in Section 3.6 to try to determine the local homeomorphism transforming the nonlinear flow (eq. (3.26)) into a linear one (eq. (3.29)). We know that in principle this can be achieved when the linearized operator $\mathscr{L}$ has no zero or purely imaginary eigenvalues, but for the time being we ignore this condition and try to perform the reduction process explicitly. We expect that at some stage of the procedure the difficulties, if any, arising from critical eigenvalues will show up and impose certain natural restrictions.

We begin by performing on the canonical form of eqs. (3.26) a linear change of variable (eq. (4.11a)) transforming the linearized operator $\mathscr{L}$ into a diagonal form $\mathbf{A}$. We ignore at this stage the complications arising from eigenvalues of algebraic multiplicity strictly larger than the geometrical one and assume from now on that the Jordan blocks of $\mathbf{A}$ are of order 1. Eqs. (3.26) become, in these new variables $\mathbf{y}$,

$$\mathrm{d}\mathbf{y}/\mathrm{d}t = \mathbf{A}\mathbf{y} + \mathbf{v}(\mathbf{y}) \qquad (5.69)$$

where $\mathbf{v}$ is the transform of $\mathbf{h}$ under the action of the diagonalizing transformation. In many problems of interest $\mathbf{v}$ will be a formal power series of $\mathbf{y}$.

We now attempt to eliminate from $\mathbf{v}(\mathbf{y})$ as many nonlinearities as possible by means of a change of variables close to the identity,

$$\mathbf{y} = \mathbf{z} + \boldsymbol{\phi}(\mathbf{z}) \qquad (5.70)$$

where $\phi(\mathbf{z})$ (to be determined by the process) is a vector polynomial of order $r \geq 2$, that is to say a vector whose components are polynomials. A vector polynomial can also be viewed as the sum of vector monomials, that is vector polynomials having one nontrivial monomial component (e.g. $z_1^{m_1} \dots z_n^{m_n}$) and all other components equal to zero; the number $m = m_1 + \dots m_n$ will be the order of the monomial.

Substituting (5.70) into (5.69) one obtains

$$\frac{d\mathbf{z}}{dt} + \frac{\partial \phi}{\partial \mathbf{z}}\frac{d\mathbf{z}}{dt} = \mathbf{A}\mathbf{z} + \mathbf{A}\phi + \mathbf{v}[\mathbf{z} + \phi(\mathbf{z})] \qquad (5.71a)$$

To eliminate the lowest nonlinearity of $\mathbf{v}$ (say of order r) it suffices to require that the following relation be satisfied:

$$\mathbf{L_A} \cdot \phi = \frac{\partial \phi}{\partial \mathbf{z}} \cdot (\mathbf{A} \cdot \mathbf{z}) - \mathbf{A} \cdot \phi = \mathbf{v}(\mathbf{z}) \qquad (5.72)$$

in which case (5.71) will reduce to

$$d\mathbf{z}/dt = \mathbf{A} \cdot \mathbf{z} + \text{(terms of order higher than } r) \qquad (5.71b)$$

The central point in the process is therefore to solve eq. (5.72), known as the *homological equation*, for the transformation function ϕ. To this end we shall now open a digression and study the principal properties of the homological operator $\mathbf{L_A}$, eq. (5.72).

Let $\omega_1, \dots, \omega_n$ and $\mathbf{u}_1, \dots, \mathbf{u}_n$ be the eigenvalues and eigenvectors of $\mathbf{A}$. We shall show that (Arnol'd, 1980):

the eigenvectors of $\mathbf{L_A}$ are vector monomials of the form $z^m \mathbf{u}_s$, where $z^m = z_1^{m_1} \dots z_n^{m_n}$

the eigenvalues of $\mathbf{L_A}$ are linear combinations of the $\{\omega_m\}$

To see this we evaluate the action of the two parts of $\mathbf{L_A}$ on the vector monomial

$$z^m \mathbf{u}_s = z_1^{m_1} \cdots z_n^{m_n} \mathbf{u}_s \qquad (5.73a)$$

We have, noticing that $\mathbf{u}_s$ is an eigenvector of $\mathbf{A}$,

$$\mathbf{A} \cdot z^m \mathbf{u}_s = z_1^{m_1} \cdots z_n^{m_n} \omega_s \mathbf{u}_s = \omega_s z^m \mathbf{u}_s \qquad (5.73b)$$

Furthermore, expressing explicitly the scalar products in the first part of $\mathbf{L_A}$ we have:

$$\frac{\partial z^m \mathbf{u}_s}{\partial \mathbf{z}} \cdot (\mathbf{A} \cdot \mathbf{z}) = \sum_{jk}\left(A_{jk} z_k \frac{\partial}{\partial z_j} z^m \mathbf{u}_s\right)$$

or, remembering that $\mathbf{A}$ is diagonal $(A_{jk} = \omega_j \delta_{jk}^{kr})$

$$\frac{\partial z^m \mathbf{u}_s}{\partial \mathbf{z}} \cdot (\mathbf{A} \cdot \mathbf{z}) = \sum_j m_j \omega_j z^m \mathbf{u}_s \tag{5.73c}$$

Combining (5.73b) and (5.73c) we arrive at

$$\mathbf{L_A} \cdot z^m \mathbf{u}_s = \left(\sum_j m_j \omega_j - \omega_s \right) z^m \mathbf{u}_s \tag{5.74}$$

which proves the statement about the eigenvalue problem of $\mathbf{L_A}$.

We come back now to the solution of eqs. (5.72). We express both the unknown vector ϕ and the (known) right hand side as sums of vector monomials,

$$\left. \begin{array}{l} \mathbf{v} = \sum_{s, m} v_{m, s} \, z^m \mathbf{u}_s \\[10pt] \phi = \sum_{s, m} h_{m, s} \, z^m \mathbf{u}_s \end{array} \right\} \tag{5.75}$$

Substituting into the equation and identifying the coefficients of equal powers in each of the basis vectors $\mathbf{u}_s$ we obtain

$$h_{m, s} = \frac{v_{m, s}}{\sum m_i \omega_i - \omega_s} \tag{5.76}$$

provided that the denominators are not equal to zero. This condition is violated when one of the eigenvalues can be expressed as a linear combination (with integer coefficients) of at least two eigenvalues,

$$\omega_s = \sum_i m_i \omega_i, \quad \sum_j |m_j| \geq 2 \tag{5.77}$$

Now, this is precisely what happens in the vicinity of a bifurcation point. For instance, in a Hopf bifurcation one has a pair of complex conjugate eigenvalues ω_1, ω_2 with vanishing real parts,

$$\omega_1 + \omega_2 = 0$$

or

$$\left. \begin{array}{l} \\[6pt] \omega_1 = 2\omega_1 + \omega_2 \end{array} \right\} \tag{5.78}$$

We refer to these situations as *resonances* – for instance, eq. (5.78) defines a resonance of order three. The point is that in the presence of resonance eq. (5.76) does not make sense and the linearization procedure fails.

One way out of this difficulty is to give up the idea of eliminating all the nonlinearities and to limit oneself only to the ones that are not associated with resonance. In practical terms we include in eq. (5.72) only the nonresonant part of $\mathbf{v}(\mathbf{z})$ and augment consequently eq. (5.71b) with the missing resonant part,

$$\mathrm{d}\mathbf{z}/\mathrm{d}t = \mathbf{A} \cdot \mathbf{z} + \sum_r \mathbf{w}_r(\mathbf{z}) \tag{5.79a}$$

where $\mathbf{w}_r(\mathbf{z})$ is a sum of resonant monomials,

$$\mathbf{w}_r(\mathbf{z}) = \sum_{s,m} w_{r,sm} z^m \mathbf{u}_s \tag{5.79b}$$

with

$$\omega_s = \sum_i m_i \omega_i \tag{5.79c}$$

As an example, in the presence of Hopf bifurcation, the unique resonant monomial has the form $z_1^2 z_2$ or, since $z_2 = z_1^*$, $|z_1|^2 \cdot z_1$. This is, precisely, the nonlinearity that we found in the perturbative approach of Section 5.7 (eq. (5.53)).

In practice the computation of the normal form is carried out exactly at the bifurcation (resonance) point. An appropriate unfolding exhibiting the system's parameters must be performed subsequently in the vicinity of this point. As mentioned in the previous section the question of the universality of the unfolding is still an open problem for most of the high codimension bifurcations.

The explicit construction of the normal form by the above method for the Brusselator model (eqs. (2.44)) and for two coupled Brusselators has been carried out by Wang (1983) and Wang and Nicolis (1987). In most cases of interest the calculations are long and tedious. Fortunately the advent of symbolic calculus has opened new possibilities. Normal form calculations that would have been very hard to imagine just a few years ago can now be performed on a routine basis (Rand and Armbruster, 1987).

An elegant variant of classical normal form theory appealing to geometrical ideas related to the center manifold has been developed by Arneodo, Coullet and coauthors (Arnéodo and Thual, 1988; Coullet and Spiegel, 1983). One seeks for solutions of eqs. (3.26) of the form

$$\mathbf{x} = \sum_i z_i(t)\mathbf{u}_{c,i} + \sum_j w_j(t)\mathbf{u}_{nc,j} \tag{5.80}$$

where $\mathbf{u}_{c,i}$ are the *critical eigenvectors* of the linearized operator $\mathscr{L}$ in the sense that they are associated with the eigenvalues which have vanishing real parts at criticality, and $\mathbf{u}_{nc,j}$ are the remaining noncritical eigenvectors. The amplitudes z_i and w_j constitute a new set of variables. Their equations of evolution are obtained by substituting (5.80) into (3.26) and by using the linear independence of $\{\mathbf{u}_i\}$. One obtains:

$$
\left.
\begin{aligned}
dz_i/dt &= \sum_{i'} \mathbf{A}_{ii'} z_{i'} + f_i(\{z_i\}, \{w_j\}) & i = 1, \ldots, d \\[2mm]
dw_j/dt &= \sum_{j'} M_{jj'} w_{j'} + g_j(\{z_i\}, \{w_j\}) & j = d+1, \ldots, n
\end{aligned}
\right\}
\quad (5.81)
$$

d being the total number of critical modes.

From the second set of eqs. (5.81) it is clear that at or near criticality the w_j vary on a time scale that is much faster than the time scale of z_i, since the eigenvalues of $\{M_{jj'}\}$ have a finite real part. It must therefore be possible to perform some kind of *adiabatic elimination*,

$$
w_j = W_j(\{z_i\}) \tag{5.82}
$$

which upon substitution into the first set of eqs. (5.81) gives a closed set of equations for $\{z_i\}$,

$$
dz_i/dt = \sum_{i'} K_{ii'} z_{i'} + q_i(\{z_i\}) \tag{5.83}
$$

Eqs. (5.82) can be regarded as the equations of the *center manifold* on which the relevant part of the dynamics lies (Guckenheimer and Holmes, 1983), whereas eqs. (5.83) should be equivalent to the normal form equations. The explicit construction of W_j and q_i can be carried out by perturbing around the critical values of the control parameters and involves, as one might have expected, solvability conditions ensuring the invertibility of the homological operator. In many instances the procedure can be substantially simplified by invoking the symmetries satisfied by the underlying system.

We end this section by pointing out that, in addition to the distance to criticality, there exists a second general mechanism ensuring the reduction of a multivariate system into a low order dynamics. Indeed, in many cases physical systems away from any sort of criticality may still give rise to widely separated time scales due to order of magnitude differences in the values of parameters and/or the state variables. For instance, in chemistry a catalytic reaction under laboratory conditions usually involves catalyst concentrations that are much less than the initial or final product concentrations, and rates that are much higher than the intrinsic rates in the absence of catalyst. As a result, some intermediate steps involving catalytic complexes proceed very quickly. In combustion the activation energies of some of the exothermic reactions are very high so these reactions proceed, at least in the early stages, much more slowly than does energy transport. Similar examples can be found in hydrodynamics, optics, biology and other fields.

To formulate this problem quantitatively one performs an appropriate scaling of variables and parameters casting the initial set of equations into the form

$$dX/dt = F(X, Y, \varepsilon) \quad \text{(slow variables)}$$
$$\varepsilon \, dY/dt = G(X, Y, \varepsilon) \quad \text{(fast variables)}$$
$$\left.\right\} \quad (5.84)$$

where $\varepsilon \ll 1$ accounts for the difference between scales. Tikhonov (Wasow, 1965) has spelled out the conditions under which from the second set of eqs. (5.84) the variables Y can be eliminated in the limit $\varepsilon \to 0$ in favor of X,

$$Y = W(X) \qquad (5.85a)$$

thereby reducing the problem to a closed set of equations for the slow variables,

$$dX/dt = F(X, W(X))$$
$$= f(X) \qquad (5.85b)$$

The 'equation of state' (5.85a) defines the *slow manifold* present in the problem. This manifold carries the relevant part of the dynamics and plays therefore a role analogous to the center manifold of bifurcation theory. An illustration of the reduction process to this manifold is provided by the ideas of a rate-limiting step and of a quasi-steady-state approximation familiar in chemistry and, most particularly, in enzyme kinetics; or by the overdamped harmonic oscillator, where Newton's equation of motion reduces to a first order equation in which inertia is discarded. For further illustrations of this important point we refer the reader to Problem 5.10.

Problems

5.1 Consider the dynamical system

$$dx/dt = \lambda x + y^3$$
$$dy/dt = -x^3 + \lambda y$$

(*a*) Compute the fixed points and their linear stability. (*b*) Study the eigenvalue problem of the linearized operator in the limit $\lambda \to 0$. Is $\lambda = 0$ a bifurcation point and if not, which among the conditions of theorem 1 of Section 5.1 is not fulfilled?

5.2 The Lorenz model (eqs. (3.4)) gives rise to a pitchfork bifurcation at

$r = 1$. Derive the normal form equation in the vicinity of this point, compute the bifurcating branches and check their stability.

5.3 Show that in the hoop problem (Sections 1.2 and 2.1) the new equilibria beyond ω_c appear through a pitchfork bifurcation. Derive the normal form equation of this bifurcation and compare with eq. (2.9).

5.4 In the region of three fixed points the Schlögl second model (eq. (2.45b)) and the Semenov model (eq. (2.47)) exhibit two limit points in the variable versus control parameter diagram. Derive the explicit values of the coefficients of the normal form equation (5.36) around these points.

5.5 The Saltzman model (Problem 4.8) generates a Hopf bifurcation for the parameter values $b = 1$, $a > b$. Using the procedure outlined in Section 5.7 and Appendix A2 derive the normal form equation in the vicinity of $b = 1$.

5.6 Compute the amplitude of the oscillations of the variables X and Y and the correction to the critical frequency $\Omega_c = A$ in the Brusselator model slightly above its Hopf bifurcation point.

5.7 Show that the coefficient P_1 in the normal form of the Hopf bifurcation (eq. (5.33)) is still given by eq. (5.25). This coefficient happens to be real in the Brusselator model. Is it also real in the Saltzman model?

5.8 Consider the dynamical system

$$dx/dt = \lambda x + \omega y + xz$$

$$dy/dt = -\omega x + \lambda y + yz$$

$$dz/dt = \sigma z - (x^2 + y^2 + z^2)$$

Show that $\sigma = -2\lambda$ is a bifurcation point of quasi-periodic solutions emerging from the reference state $(x_s(t), y_s(t), z_s)$ where x_s, y_s are periodic and z_s time-independent. Hint: Switch to cylindrical coordinates (Langford, 1979).

5.9 Rederive the normal form equations for the Brusselator and Saltzman models using the procedure of Section 5.9 (Wang and Nicolis, 1987).

5.10 The rate equations for the Oregonator (eq. (2.43)) read, after appropriate scaling transformations of variables and parameters (Tyson, 1976)

$$dx/d\tau = s(y - xy + x - qx^2)$$

$$dy/d\tau = \frac{1}{s}(-y - xy + fz)$$

$$dz/d\tau = w(x - z)$$

where $s \gg 1$ and f, q, w are of the order of or less than unity. (*a*) Derive the conditions of Hopf bifurcation using f as the control parameter. (b) Using Tikhonov's theorem (Section 5.9) perform the adiabatic elimination of variable x and check whether the results of (a) subsist in the reduced set of equations for y and z.

CHAPTER SIX

Spatially distributed systems, broken symmetries, pattern formation

6.1 General formulation

As we saw in Section 3.1 and throughout Chapter 2 the macroscopic description of systems composed of many particles gives rise, typically, to state variables that are *fields* in the sense that they depend continuously on the space coordinates. The evolution laws of these variables are expressed as partial differential equations of the form

$$\partial X_i(\mathbf{r}, t)/\partial t = F_i(\{X_j(\mathbf{r}, t)\}, \{\nabla^k X_j(\mathbf{r}, t)\}, \lambda) \tag{6.1}$$

They involve spatial derivatives like the Laplacian of temperature $\nabla^2 T$ or concentration $\nabla^2 c$, or the gradient of the velocity $\nabla \mathbf{v}$ and include as particular cases the equations of fluid dynamics and the reaction–diffusion equations. A dynamical system of the form (6.1), hereafter referred to as the *spatially distributed system* possesses, in principle, an infinity of variables – the values of the fields $\{X_i(\mathbf{r}, t)\}$ at each point in space coupled through the transport phenomena generated by the spatial inhomogeneities. As shown in Section 3.1, in some cases this complication can be bypassed by a Galerkin expansion (eq. (3.2)) truncated to the first few modes. The present chapter is devoted to problems where this reduction is not applicable and spatial dependences need to be incorporated explicitly in the description. We shall focus more specifically on the new features arising from the presence of spatial degrees of freedom.

The ideas and techniques developed in the last three chapters suggest the following canonical procedure for studying eqs. (6.1).

(i)   First a suitable reference state $\mathbf{X}_s$, which is an exact solution of eqs. (6.1), is identified. The choice of $\mathbf{X}_s$ is motivated by physical arguments. Typically $\mathbf{X}_s$ is a state describing the 'simplest' behavior observed in the system – for instance, the state of rest in the thermal convection (Bénard) problem (Section 1.3), or the uniform steady state in a chemically reacting system (Section 1.4).

(ii) The next step is to test the stability of $\mathbf{X}_s$ against perturbations. To reduce the initial nonlinear problem to a linear one, one needs to extend the principle of linearized stability of Section 3.6. This extension, which amounts essentially to redefining an appropriate 'distance' in an infinitely-dimensional space, has been carried out in a number of cases (Sattinger, 1972) and will be taken for granted from now on. In the presence of spatial inhomogeneities the linear stability operator replacing the operator $\mathscr{L}$ in eqs. (3.29) is expected to contain spatial derivatives. As a result the eigenvectors will span an infinite-dimensional functional space and there will be an infinity of eigenvalues associated with them. As we shall see in the subsequent sections the most important new element arising at the level of stability analysis from the presence of spatial derivatives in the evolution laws is the possibility of *spontaneous symmetry breaking instabilities*, that is to say, instabilities with respect to space-dependent perturbations which are less symmetric than the evolution laws or the reference state $\mathbf{X}_s$.

(iii) Finally, the nonlinear behavior around $\mathbf{X}_s$ beyond the instability point is explored by a perturbation method closely following Section 5.2, the aim being to obtain normal form equations for the amplitude of the solutions to the dominant order. As we shall see the form of these equations will depend crucially on the *size* of the system. In systems of small spatial extent much of the analysis of the previous chapter will extend straightforwardly. But in systems of large spatial extent some new features will appear, due to the interaction between the spatial degrees of freedom.

Spatially distributed systems described by state variables in the form of fields also arise in the microscopic description of the electromagnetic field and of its coupling with matter as well as in elementary particle physics. We do not address this type of problem here. It is worth pointing out, however, that the origin of many of the symmetry-breaking phenomena which arise in field theory and particle physics is quite analogous to that arising in the macroscopic physics of spatially distributed systems.

6.2 The Bénard problem: reference state and linearization of the Boussinesq equations

We shall apply the above described procedure to the problem of the thermal convection instability, whose quantitative formulation in the

Boussinesq approximation was laid down in Section 2.5. We choose as the reference solution the (stationary) state at rest $\mathbf{v} = 0$ for which eqs. (2.33) and (2.35) reduce to (cf. also eq. (2.32))

$$\nabla p_s = -\rho_s g \mathbf{1}_z$$

$$= -\rho_0(1 - \alpha(T_s - T_0))g\mathbf{1}_z \tag{6.2}$$

$$\nabla^2 T_s = 0 \tag{6.3}$$

In view of the geometry of the experiment and the symmetries involved in the problem we expect that p and T vary only along the z direction. Integration of eq. (6.3) using the boundary conditions (2.36) leads then to

$$T_s(z) = T_0 - \beta z$$

with

$$\beta = \frac{T_0 - T_1}{d} \tag{6.4}$$

Substituting into eq. (6.2) and integrating once again with respect to z one finds

$$p_s(z) = p_0 - \rho_0 g\left(1 + \frac{\alpha\beta z}{2}\right)z \tag{6.5}$$

Notice that the characteristics of the system in this state are independent of the kinetic coefficients η and λ which appear in the full balance equations.

We now study the effect of perturbations on the system around the above reference state $(T_s, p_s, \rho_s, \mathbf{v}_s = 0)$. Specifically, we set

$$\left.\begin{array}{l} T = T_s(z) + \theta(\mathbf{r}, t) \\[4pt] \rho = \rho_s(z) + \delta\rho(\mathbf{r}, t) \\[4pt] p = p_s(z) + \delta p(\mathbf{r}, t) \\[4pt] \mathbf{v} = \delta\mathbf{v}(\mathbf{r}, t): \qquad \delta\mathbf{v} = (u, v, w) \end{array}\right\} \tag{6.6}$$

and linearize the equations of Section 2.5 with respect to θ, $\delta\rho$, δp, $\delta\mathbf{v}$. Since $\delta\mathbf{v}$ is a first order term it can only be multiplied by ∇T_s in eq. (2.35). The linearized version of this equation is therefore, using eq. (6.4):

$$\rho_0 c_p\left(\frac{\partial\theta}{\partial t} - \beta w\right) = \lambda\nabla^2\theta$$

or, introducing the *thermal diffusivity* coefficient

$$\kappa = \lambda/\rho_0 c_p \tag{6.7}$$

$$\partial\theta/\partial t = \beta w + \kappa\nabla^2\theta \qquad (6.8)$$

We consider next the Navier–Stokes equation (2.33). Substituting (6.6) and taking the reference state relations (6.4) and (6.5) into account we obtain

$$\frac{\partial\delta\mathbf{v}}{\partial t} = -\frac{1}{\rho_0}\nabla\delta p + \frac{\eta}{\rho_0}\nabla^2\delta\mathbf{v} - \frac{g}{\rho_0}\delta\rho\mathbf{1}_z$$

Introducing the kinematic viscosity

$$v = \eta/\rho_0 \qquad (6.9)$$

and evaluating $\delta\rho$ from the equation of state (2.32), $\delta\rho = -\rho_0\alpha\theta$, we obtain the more convenient form

$$\frac{\partial\delta\mathbf{v}}{\partial t} = -\frac{1}{\rho_0}\nabla\delta p + v\nabla^2\delta\mathbf{v} + g\alpha\theta\mathbf{1}_z \qquad (6.10)$$

Eqs. (6.8) and (6.10) constitute the basic equations of linear stability analysis of the Bénard problem. The introduction of the kinetic coefficients κ and v instead of the initial ones λ and η is doubly advantageous. First, κ and v vary less sensitively with the material considered. Second, the parameters ρ_0, c_p are absorbed in the definition of these coefficients (ρ_0 still subsists in the pressure term, but this is immaterial as we shall see shortly). Still, eqs. (6.8) and (6.10) contain no less than five control parameters $(\beta, \kappa, v, g\alpha, d)$, d being introduced by the boundary conditions. Since the objective of stability theory is to identify the values of control parameters at which a qualitative change of behavior takes place, it is obviously advantageous to get rid of as many spurious parameters as possible. A powerful first step in this direction is to express the equations in *dimensionless form*, the additional advantage of this being to allow for comparison of experiments obtained under different conditions. In the present problem the following dimensional quantities are involved:

For the independent variables:
  time t, $[T]$
  space $\mathbf{r}$, $[L]$
For the dependent variables:
  temperature θ, $[K]$
  velocity $\delta\mathbf{v}$, $[LT^{-1}]$
  pressure δp, $[ML^{-1}T^{-2}]$

To switch to dimensionless form we seek combinations of parameters present in our problem that have the above dimensions. This leads to the following scaling:

$$r_i' = r_i/d$$

$$t' = t/(d^2/\kappa)$$

$$\theta' = \theta/(\nu\kappa/g\alpha d^3) \qquad\qquad \left.\right\} \quad (6.11)$$

$$\delta v_i' = \delta v_i/(\kappa/d)$$

$$\delta p' = \delta p/(\rho_0 \nu\kappa/d^2)$$

Substituting (6.11) into (6.8) and (6.10) to which we also add the incompressibility condition (2.28) and suppressing the primes to simplify notation we obtain the following dimensionless form of the linearized Boussinesq equations:

$$\partial\theta/\partial t = Rw + \nabla^2\theta$$

$$\partial\delta\mathbf{v}/\partial t = P\{ -\nabla\delta p + \nabla^2\delta\mathbf{v} + \theta\mathbf{1}_z\} \qquad \left.\right\} \quad (6.12)$$

$$\operatorname{div}\delta\mathbf{v} = 0$$

where we introduced the dimensionless parameters

$$R = \alpha g\beta d^4/\nu\kappa = ag\Delta Td^3/\nu\kappa \qquad \text{(Rayleigh number)} \qquad \left.\right\} \quad (6.13)$$

$$P = \nu/\kappa \qquad \text{(Prandtl number)}$$

We have thus reduced the number of parameters to two. An additional advantage of the reduction to dimensionless form is that one of the new parameters, the Rayleigh number, combines in an elegant and compact way the factors at the origin of the instability (the numerator of the first eq. (6.13)) and the stabilizing factors (the denominator).

One may simplify eqs. (6.12) further by reducing them to a closed form for $(\theta, \delta\mathbf{v})$. The elimination of δp can be achieved by taking the curl of both sides of the second relation (6.12). Introducing the *vorticity* $\boldsymbol{\omega}$,

$$\boldsymbol{\omega} = \nabla \times \delta\mathbf{v} \qquad (6.14)$$

we obtain

$$\frac{\partial\boldsymbol{\omega}}{\partial t} = P\left[\left(\mathbf{1}_x\frac{\partial\theta}{\partial y} - \mathbf{1}_y\frac{\partial\theta}{\partial x}\right) + \nabla^2\boldsymbol{\omega}\right] \qquad (6.15)$$

Taking once more the curl of this latter relation and using the vector identity $\nabla \times (\nabla \times \mathbf{a}) = \nabla(\operatorname{div}\mathbf{a}) - \nabla^2\mathbf{a}$ we obtain, using the first relation (6.12):

$$\frac{\partial}{\partial t}\nabla^2\delta\mathbf{v} = P\left(-\nabla\frac{\partial\theta}{\partial z} + \nabla^2\theta\mathbf{1}_z + \nabla^4\delta\mathbf{v}\right) \qquad (6.16)$$

Projecting this equation on the z-axis we finally obtain

$$\frac{\partial}{\partial t} \nabla^2 w = P\left(\frac{\partial^2\theta}{\partial x^2} + \frac{\partial^2\theta}{\partial y^2} + \nabla^4 w\right) \tag{6.17}$$

Together with the first relation of (6.12) this equation constitutes a closed set of two equations for (θ, w). Their analysis, which will provide us with the solution of the stability problem, is carried out in the next section.

6.3 The Bénard problem: linear stability analysis for free boundaries

Eqs. (6.12) and (6.17) constitute a set of linear homogeneous equations of first order in time with constant coefficients. They therefore admit solutions of the form (cf. eq. (4.1)):

$$\begin{pmatrix} \theta(\mathbf{r}, t) \\ w(\mathbf{r}, t) \end{pmatrix} = e^{\omega t}\begin{pmatrix} \tilde{\theta}(\mathbf{r}) \\ \tilde{w}(\mathbf{r}) \end{pmatrix} \tag{6.18}$$

Substituting back to the original equations one finds

$$\left. \begin{aligned} \omega\tilde{\theta} &= \nabla^2\tilde{\theta} + R\tilde{w} \\[2mm] \frac{\omega}{P}\nabla^2\tilde{w} &= \frac{\partial^2\tilde{\theta}}{\partial x^2} + \frac{\partial^2\tilde{\theta}}{\partial y^2} + \nabla^4\tilde{w} \end{aligned} \right\} \tag{6.19}$$

We notice that in the right hand side spatial derivatives appear through operators containing $\partial^2/\partial x^2 + \partial^2/\partial y^2$ and $\partial^2/\partial z^2$ in an additive fashion. Eqs. (6.19) admit therefore solutions of the form

$$\begin{pmatrix} \tilde{\theta} \\ \tilde{w} \end{pmatrix} = \begin{pmatrix} \theta_{kn} \\ w_{kn} \end{pmatrix}\phi_k(x, y)\psi_n(z) \tag{6.20}$$

where θ_{kn}, w_{kn} are space-independent and ϕ_k, ψ_n are, respectively, eigenfunctions of the above two operators.

To solve the eigenvalue problem for the 'transverse Laplacian' $\nabla_t^2 = \partial^2/\partial x^2 + \partial^2/\partial y^2$ one notices that, in the limit of a large aspect ratio (cf. Section 1.2), the system can be treated as being unbounded in the horizontal direction. Using familiar arguments similar to those applied in wave mechanics one realizes that such systems can be modeled adequately by *periodic boundary conditions*, in which case

$$\phi_k(x, y) = e^{i(k_x x + k_y y)} \tag{6.21a}$$

The corresponding eigenvalue, obtained by acting on (6.21a) by ∇_t^2, is then

$$\lambda_k = -k^2 = -(k_x^2 + k_y^2) \tag{6.21b}$$

where the periodicity of boundary conditions is ensured by the property

$$k_x = \frac{2\pi}{L_x} n_x, \qquad k_y = \frac{2\pi}{L_y} n_y \qquad (6.21c)$$

Here L_x, L_y are the system size along the x and y directions and n_x, n_y take integer values.

Turning now to the eigenvalue problem of d^2/dz^2, we want to solve

$$d^2\psi_n/dz^2 = \mu_n\psi_n$$

subject to the boundary conditions (2.36) and (2.37) and to (2.38) and (2.39) or (2.40)–(2.42). Since we are dealing here with the perturbations rather than the initial variables, eqs. (2.36) and (2.37) reduce to (remembering also that we have rescaled our variables and parameters)

$$\psi_n(0) = \psi_n(1) = 0 \qquad (6.22a)$$

For simplicity we shall also choose the free boundary conditions,

$$(d^2\psi_n/dz^2)_0 = (d^2\psi_n/dz^2)_1 = 0 \qquad (6.22b)$$

Under these conditions the solution of the eigenvalue equation for d^2/dz^2 becomes

$$\psi_n = \sin n\pi z, \qquad 0 \le z \le 1 \qquad (6.23a)$$

$$\mu_n = -n^2\pi^2 \qquad (6.23b)$$

where n is an integer. Substituting into (6.20) we obtain $\tilde{\theta}$ and $\tilde{w}$ in the form

$$\begin{pmatrix} \tilde{\theta} \\ \tilde{w} \end{pmatrix} = \begin{pmatrix} \theta_{kn} \\ w_{kn} \end{pmatrix} e^{i(k_x x + k_y y)} \sin n\pi z \qquad (6.24)$$

Since k_x, k_y, and n determine the wavelength of the perturbations along the x, y and z directions, the amplitudes θ_{kn}, w_{kn} may be interpreted as the coefficients of expansion of $\tilde{\theta}, \tilde{w}$ in Fourier series. We refer to them as *normal modes*. Substituting further eq. (6.24) into eqs. (6.19) we see that space dependences are factored out on both sides and we obtain a set of purely algebraic equations for the normal modes, which for comparison with Chapter 4 we write in the suggestive form

$$\begin{pmatrix} -(k^2 + n^2\pi^2) - \omega & R \\ -k^2 & (k^2 + n^2\pi^2)^2 + \dfrac{\omega}{P}(k^2 + n^2\pi^2) \end{pmatrix} \begin{pmatrix} \theta_{kn} \\ w_{kn} \end{pmatrix} = \begin{pmatrix} 0 \\ 0 \end{pmatrix}$$

$$(6.25)$$

This system of equations admits nontrivial solutions provided that the determinant of the 2×2 matrix acting on $\begin{pmatrix} \theta_{kn} \\ w_{kn} \end{pmatrix}$ vanishes. This condition is nothing but the characteristic equation (eq. (4.5)) which assumes here the explicit form

$$\frac{1}{P}(k^2 + n^2\pi^2)\omega^2 + (k^2 + n^2\pi^2)^2\left(\frac{1}{P} + 1\right)\omega + (k^2 + n^2\pi^2)^3 - Rk^2 = 0$$

(6.26)

It links the eigenvalue ω to the spatial characteristics of the perturbations ($k = (k_x^2 + k_y^2)^{1/2}$ and n) and to the control parameters R and P which take into account the properties of the material and the constraints acting on the system.

We notice from (6.26) that for given k and n the sum of eigenvalues is always negative,

$$\omega_1 + \omega_2 = -(k^2 + n^2\pi^2)(P + 1) < 0 \qquad (6.27)$$

It follows that the passage through an instability (if any) can only take place when one of the (real) roots goes through zero while the second root remains negative,

$$\left. \begin{aligned} \omega_1(R_c, P_c) &= 0 \\ \omega_2(R_c, P_c) &< 0 \end{aligned} \right\} \quad (6.28)$$

This is known in the literature of hydrodynamics as the *principle of exchange of stability* (Chandrasekhar, 1961). In this marginal state the characteristic equation (6.26) reduces to

$$(k^2 + n^2\pi^2)^3 - Rk^2 = 0 \qquad (6.29)$$

which no longer involves the Prandtl number. In the spirit of linear stability analysis (Chapter 4) we interpret this equation as a relation between the control parameter R and the characteristics of the perturbation (k, n):

$$R = R(k, n) = (k^2 + n^2\pi^2)^3/k^2 \qquad (6.30)$$

If R exceeds the value of $R(k, n)$ given by the above equation the product of the roots in eq. (6.26) will become negative. This corresponds to the situation

$$\omega_1 > 0, \qquad \omega_2 < 0$$

in other words for $R > R(k, n)$ the reference state is unstable, behaving as a (generalized) saddle. On the other hand for $R < R(k, n)$ the product of the roots ω_1, ω_2 is positive and the additional condition $\omega_1 + \omega_2 < 0$ (eq.

(6.27)) allows us to conclude that both ω_1 and ω_2 are negative and hence that the reference state is asymptotically stable.

Suppose now that one performs a Bénard experiment in which the fluid is initially at uniform temperature ($R = 0$, eq. (6.13)). By gradually increasing the temperature gradient R itself increases from values $R < R(k, n)$ to values tending to $R(k, n)$. The set of values of R for which the system will first penetrate into the domain of instability is, obviously, given by curve (6.30) with $n = 1$. The resulting function $R(k)$ is depicted in Fig. 6.1. It possesses a minimum at

$$k_c = \pi/\sqrt{2} \tag{6.31a}$$

or, reestablishing the initial (dimensional) variables through (6.11):

$$k_c = \pi/(\sqrt{2}d) \tag{6.31b}$$

The corresponding value R_c of the Rayleigh number

$$R_c = \frac{27}{4}\pi^4 \approx 657.51 \tag{6.31c}$$

gives the frontier between asymptotic stability ($R < R_c$) and instability ($R > R_c$). It is remarkable that this transition is mediated by a single universal parameter rather than by parameters reflecting the detailed structure of the material or of the experimental setup.

A most important point in connection with eqs. (6.31) and Fig. 6.1 is that R_c corresponds to nontrivial values of k and n, that is to say, to critical perturbations having a finite characteristic wavelength in the x, y and z directions given by

Fig. 6.1 Marginal stability curve $R(k)$ (eq. (6.30) with $n = 1$) separating the domain of asymptotic stability from the domain of instability of the state of rest in the Bénard problem with free–free boundary conditions.

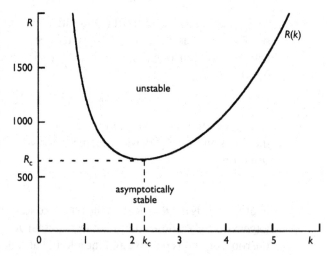

$$l_{x_c} = \frac{2\pi}{k_{x_c}}, \qquad l_{y_c} = \frac{2\pi}{k_{y_c}}, \qquad l_{z_c} = 2d \qquad (6.32)$$

with $(k_{x_c}^2 + k_{y_c}^2)^{1/2} = k_c = \pi/\sqrt{2}d$. It follows that the dominant mode in the vicinity of the instability will now be spatially inhomogeneous. This is the signature, at the level of stability analysis, of the *symmetry breaking* character of the transition. We thus have a simple elegant explanation of a large body of the experimental facts on pattern formation surveyed in Chapter 1. We notice that in the present problem the characteristic wavelength is *extrinsic*, in the sense that it is directly proportional to the depth of the layer. We defer further comments on this point to Section 6.4 dealing with chemical symmetry-breaking instabilities.

It is instructive to explore the vicinity of the critical state (k_c, R_c). By continuity we expect that ω_2 will remain a finite negative number while $|\omega_1|$ will be a small number. We may compute it by neglecting ω^2 in eq. (6.26) and by expanding systematically around the critical state, taking eqs. (6.30)–(6.31) into account. To the first nontrivial order a straightforward algebra leads to

$$\omega \approx a(R - R_c) - b(k - k_c)^2 \qquad (6.33a)$$

where the positive numbers a and b are given by

$$\left.\begin{array}{l} a = 2P/9\pi^2(P + 1) \\[2mm] b = 4P/(P + 1) \end{array}\right\} \qquad (6.33b)$$

Fig. 6.2 depicts the *dispersion relation* (6.33a) for various values of R. We notice that all the curves have a maximum at $k = k_c$. For $R = R_c$ the $\omega = \omega(k)$ curve is tangent to the k-axis, yielding $\omega_c = 0$ as expected. For

Fig. 6.2 Graph of ω versus k in the Bénard problem close to the instability threshold $R = R_c$ (eq. (6.33a)).

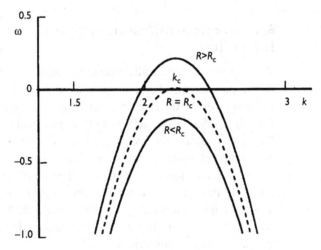

Table 6.1

	R_c	k_c
free–free	657.51	2.221
rigid–free	1100.65	2.681
rigid–rigid	1708	3.117

$R > R_c$ part of the curve lies above the k-axis, implying that there is a zone of unstable modes of which the fastest growing is the one at $k = k_c$. Now, when the lateral extension L of the system gets very large the spectrum of ks becomes continuous according to (6.21c). The number of unstable modes for $R > R_c$, however small $|R - R_c|$ might be, tends then to infinity. This is typical of most of the instabilities encountered in spatially distributed systems of large extent.

Notice that for values of k significantly different from k_c the left part of the dispersion curve crosses the negative ω-axis and subsequently continues in the negative k direction symmetrically with respect to this axis. This is a direct consequence of the fact that the characteristic equation (6.26) remains invariant under the transformation $k \rightarrow -k$.

The solution of the linear stability problem for other types of boundary conditions is carried out in Chandrasekhar's classical monograph (Chandrasekhar, 1961). Table 6.1 gives the critical values for the three most widely studied cases. We notice that rigid boundaries exert a stabilizing influence, as one might have guessed from the very fact that they force all three components of the velocity to vanish on the two horizontal boundaries.

6.4 Reaction–diffusion systems. The Turing instability

We turn next to chemically reacting systems and choose as the reference state a time-independent and spatially uniform distribution of chemicals in the reactor. In the experimental setup of an open reactor described in Section 1.4 this will be achieved for all practical purposes if the depth of the reactor along the feeding direction is much less than its lateral extension. Alternatively one may adopt the *pool chemical approximation* whereby in a closed reactor the initial products are in large excess. The intermediates then behave as an open subsystem during a substantial period of time. Depending on the case we shall adopt for them fixed (Dirichlet), zero flux (Neuman) or periodic (simulating an unbounded system) boundary conditions.

Let $\{X_{is}\}$ be the uniform steady state reference solution (in the case of fixed boundary conditions this implies that the boundary values of $\{X_i\}$ should be kept equal to $\{X_{is}\}$). Setting

$$X_i = X_{is} + x_i \qquad i = 1, \ldots, n \qquad (6.34)$$

substituting into eqs. (2.27) and linearizing with respect to x_i we obtain

$$\frac{\partial x_i}{\partial t} = \sum_j \left(\frac{\partial v_i}{\partial X_j}\right)_s x_j + D_i \nabla^2 x_i \qquad i = 1, \ldots, n \qquad (6.35)$$

where v_i denotes the overall rate of change of X_i due to the reactions. Since the coefficients $\partial v_i / \partial X_j$ and D_i are time-independent, eqs. (6.35) admit solutions of the form

$$\mathbf{x}(r, t) = \mathbf{u}\phi(\mathbf{r})\, e^{\omega t} \qquad (6.36)$$

where $\mathbf{u}$ is space-independent and accounts for the structure of $\mathbf{x}$ as a vector in concentration space. Now by construction $\partial v_i / \partial X_j$ are space independent, and the Laplacian is the only operator acting on space coordinates in eqs. (6.35). It suffices therefore to choose $\phi(\mathbf{r})$ to be an eigenfunction of this operator subject to the boundary conditions that apply to the problem under consideration:

$$\nabla^2 \phi_m(\mathbf{r}) = -k_m^2 \phi_m(\mathbf{r}) \qquad (6.37)$$

Here m is a set of indices labeling the (infinite) set of eigenfunctions and the minus sign in front of k_m^2 accounts for the fact that ∇^2 is a dissipative operator having nonpositive eigenvalues. Notice that $k_m = 0$ corresponds to a uniform distribution and $k_m \neq 0$ to a nonuniform one.

Substituting (6.36) and (6.37) into (6.35) one realizes that the space and time dependences are factored out on both sides of the equations. One is thus left with a homogeneous set of algebraic equations for the components u_i of the vector $\mathbf{u}$ of the form

$$\sum_j \left[\left(\frac{\partial v_i}{\partial X_j}\right)_s - (\omega + D_i k_m^2)\delta_{ij}^{kr}\right] u_j = 0 \qquad (6.38)$$

This system admits a nontrivial solution provided that the characteristic equation (eq. (4.5), see also (6.26)) is satisfied:

$$\det \left|\left(\frac{\partial v_i}{\partial X_j}\right)_s - (\omega + D_i k_m^2)\delta_{ij}^{kr}\right| = 0 \qquad (6.39)$$

In this equation it is understood that the coefficients $(\partial v_i / \partial X_j)_s$ depend on the control parameters λ present in the problem such as the temperature or the concentration of an initial chemical (like the parameter B in the

Brusselator, eq. (2.44)), with the exception of the diffusion coefficients which appear explicitly through the terms $D_i k_m^2$. In the limit where all diffusion coefficients are identical, $D_i = D$, one can absorb these terms in a redefinition of ω,

$$\omega' = \omega + D k_m^2 \tag{6.40}$$

Expressed in terms of ω', eq. (6.39) reduces, then, to the characteristic equation of a spatially uniform system. We know by now that such a system may admit bifurcations associated with one real eigenvalue crossing zero ($\omega' = 0$), a pair of complex conjugate eigenvalues crossing the real axis ($\mathrm{Re}\,\omega' = 0$, $\mathrm{Im}\,\omega' \neq 0$), or various higher codimension situations. In each case at such a transition point the eigenvalue ω of the original system will have by virtue of (6.40) a real part smaller than or equal to zero, given by $-D k_m^2$. If follows that

the first instability in a system having equal diffusion coefficients will be associated with spatially homogeneous solutions ($k_m = 0$);
if $k_m = 0$ is excluded by the boundary conditions diffusion will play a stabilizing role, postponing the instability that would take place in its absence.

Let us turn now to the general case where at least two among the diffusion coefficients are unequal. Excluding for the moment higher codimension bifurcations the following possibilities can be envisaged.

A A real eigenvalue crosses zero at criticality, $\omega_c = 0$

Eq. (6.39) provides a relation $f(\lambda, k_m^2) = 0$ between the control parameter λ contained in $(\partial v_i / \partial X_j)_s$ and the eigenvalue of the Laplacian k_m^2 which accounts for the spatial characteristics of the perturbations. This relation is of $2n$th degree in k and is invariant under the transformation $k \to -k$. Solving with respect to λ,

$$\lambda = \lambda(k_m^2) \tag{6.41}$$

one expects to find a marginal stability curve having at least one extremum. Two typical cases are depicted in curves (a) and (b) of Fig. 6.3, limited for compactness to the positive k-axis only.

In (a), as λ is varied from values corresponding to asymptotic stability (here below curve (a)) to values leading to instability (above curve (a)), the first transition will take place at λ_c' for which $k_m = 0$. This corresponds to a space-independent situation. In other words, the dominant mode in the vicinity of the first bifurcation point will be a homogeneous one and the analysis will reduce to that carried out in the previous two chapters.

The situation is very different in case (*b*). Here the extremum λ_c'' occurs at a nontrivial value $k_m = k_{m_c}$ where the dominant mode of the solutions has a nontrivial space dependence displaying a *characteristic length* $l_c \approx k_{m_c}^{-1}$. We are thus in the presence of a symmetry-breaking instability similar to the one observed in the Bénard problem (Fig. 6.1) with one most important difference: the characteristic length l_c is here completely intrinsic since it is related by (6.41) to the system's parameters independently of geometry and boundary conditions. We shall refer to this type of instability as the *Turing instability*, since it was foreseen in Turing's historic paper (Turing, 1952) on the chemical basis of morphogenesis. This type of instability allows one to understand, at least qualitatively, the experimental data described in Figs. 1.11–1.13.

It is interesting to identify the reasons why symmetry-breaking instabilities in chemistry can give rise to intrinsic characteristic lengths, whereas in hydrodynamics these lengths are always extrinsic. In chemistry one witnesses the coexistence of a transport phenomenon, whose characteristic parameter is a diffusion coefficient D of dimensions L^2T^{-1} and of a local relaxation phenomenon whose characteristic parameter is a rate constant of dimensions T^{-1}. From these one can construct a combination $l_c \approx (D/k)^{1/2}$ having the dimensions of a length. On the other hand in hydrodynamics one still has a diffusive heat or momentum transport, but one lacks a preferred characteristic time since hydrodynamic modes display a continuum of frequencies accumulating to zero.

As in the previous section, it is interesting to derive for the two cases depicted in Fig. 6.3 the dispersion relation linking ω to k in the vicinity of the criticality. In the limit of small ω all terms of (6.39) except the first and zero order ones in ω can be neglected. This gives rise to an equation of the

Fig. 6.3 Two typical outcomes of linear stability analysis around a uniform steady state in a spatially distributed reaction–diffusion type system. (*a*) The marginal stability curve $\lambda(k)$ presents an extremum at $k = 0$, implying that near the instability threshold the system will remain homogeneous in space. (*b*) The marginal stability curve $\lambda(k)$ possesses an extremum at a nontrivial value $k = k_{m_c}$, implying the emergence of solutions displaying broken symmetries beyond the instability threshold.

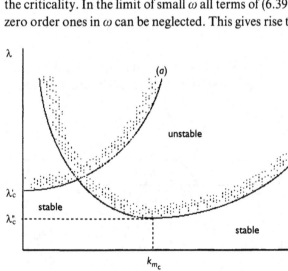

form

$$\text{(polynomial of } 2(n-1)\text{th degree in } k_m)\cdot\omega$$
$$= \text{(polynomial of } 2n\text{th degree in } k_m)$$

in which the right hand side vanishes at criticality on the grounds of eq. (6.41). Expanding (λ, k_m) locally around (λ_c, k_{m_c}) and taking into account the existence of an extremum at k_{m_c} we obtain a set of equations similar to (6.33a),

$$\omega \approx a(\lambda - \lambda_c) - bk_m^2 \qquad \text{for } k_{m_c} = 0 \quad \text{(Fig. 6.3, curve } (a)) \quad (6.42a)$$

and

$$\omega \approx a(\lambda - \lambda_c) - b(k_m - k_{m_c})^2 \qquad \text{for } k_{m_c} \neq 0 \quad \text{(Fig. 6.3, curve } (b))$$
$$(6.42b)$$

These relations are depicted in Figs. 6.4(a),(b).

A concrete illustration of eq. (6.42a), of curve (a) of Fig. 6.3 and of Fig. 6.4(a) is provided by one-variable systems such as the Schlögl model (eq. (2.45)). Of more interest is the two-variable Brusselator model, which undergoes a Turing instability. Using expressions (3.28) of the rate functions and the uniform steady state solution $(A, B/A)$ as the reference state we obtain the characteristic equation

$$\begin{vmatrix} B - 1 - (D_1 k_m^2 + \omega) & A^2 \\ -B & -A^2 - (D_2 k_m^2 + \omega) \end{vmatrix} = 0$$

or in an explicit form

$$\omega^2 - [B - A^2 - 1 - (D_1 + D_2)k_m^2]\omega$$
$$+ A^2 + [A^2 D_1 k_m^2 - (B-1)D_2 k_m^2] + D_1 D_2 k_m^4 = 0$$
$$(6.43)$$

Setting $\omega = \omega_c = 0$ at the instability threshold and using B as a control parameter we find the explicit form of eq. (6.41) as

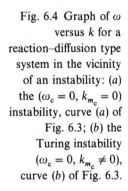

Fig. 6.4 Graph of ω versus k for a reaction–diffusion type system in the vicinity of an instability: (a) the ($\omega_c = 0$, $k_{m_c} = 0$) instability, curve (a) of Fig. 6.3; (b) the Turing instability ($\omega_c = 0$, $k_{m_c} \neq 0$), curve (b) of Fig. 6.3.

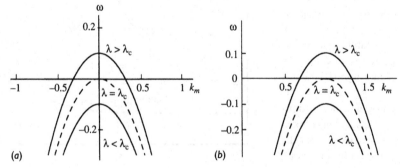

$$B = D_1 k_m^2 + A^2 \frac{D_1}{D_2} + 1 + \frac{A^2}{D_2 k_m^2} \tag{6.44}$$

This function admits a minimum

$$B_c = \left[1 + \left(\frac{D_1}{D_2} \right)^{1/2} A \right]^2 \tag{6.45a}$$

for

$$k_{m_c}^2 = A/(D_1 D_2)^{1/2} \tag{6.45b}$$

We see that, as anticipated above, B_c and k_{m_c} are intrinsic. This statement must be tempered, however, by the observation that, in principle, not all values of k_m are accessible. Indeed, taking for concreteness periodic boundary conditions, we have $k_m \sim (2\pi/L)m$ where L is the extension of the system and m is an integer. In actual fact, therefore, for L finite the critical value k_{m_c} will not be given by (6.45b) but by the closest number to this expression having the form $(2\pi/L)m$.

B A pair of complex conjugate eigenvalues crosses the imaginary axis at criticality, $\text{Re}\,\omega_c = 0, \text{Im}\,\omega_c \neq 0$

Eq. (6.39) gives rise to two equations for the real and imaginary parts of the critical eigenvalue. Setting $\text{Re}\,\omega_c = 0$, one obtains two equations linking λ, $\text{Im}\,\omega_c$ and k_m^2. Eliminating $\text{Im}\,\omega_c$ between these two equations one finds a new relation $\lambda = \lambda(k_m^2)$ replacing eq. (6.41). Finally substituting back into one of the initial relations one obtains the value of $\text{Im}\,\omega_c$ in terms of k_m and other noncritical parameters that may be present in the problem.

In principle, both of the situations depicted in Fig. 6.3 can take place here as well. The real part of the eigenvalue close to criticality will then behave as in Fig. 6.4. Since $\text{Im}\,\omega_c \neq 0$ we expect to have the onset of time-periodic solutions with an intrinsically determined period beyond the instability threshold. We may refer to this phenomenon as *time symmetry-breaking*. In Fig. 6.3, curve (*a*) and Fig. 6.4(*a*) these emerging solutions will be uniform in space, whereas in Fig. 6.3 curve (*b*) and Fig. 6.4(*b*) they will display a nontrivial space dependence associated with space symmetry-breaking. We will speak, then, of *spatio-temporal* patterns.

The first type of situation is the only one to arise in systems involving two variables. Taking again the example of the Brusselator and setting $\text{Re}\,\omega = 0$ in eq. (6.43) we find

$$-(\text{Im } \omega_c)^2 + i[B - A^2 - 1 - (D_1 + D_2)k_m^2]\text{ Im }\omega_m + A^2$$

$$+ [A^2 D_1 k_m^2 - (B-1)D_2 k_m^2] + D_1 D_2 k_m^4 = 0 \qquad (6.46)$$

The imaginary part of this equation gives a relation between k_m and the control parameter B

$$B = A^2 + 1 + (D_1 + D_2)k_m^2 \qquad (6.47a)$$

which has the form of curve (*a*) of Fig. 6.3. The minimum of this marginal stability curve is $(k_{m_c} = 0, B_c = A^2 + 1)$ which is nothing but the onset of oscillatory instability in the spatially uniform Brusselator (eq. (4.38)). On the other hand, taking the real part of (6.46) one obtains the expression of the oscillation frequency

$$(\text{Im }\omega)^2 = A^2 + [A^2 D_1 - (B-1)D_2]k_m^2 + D_1 D_2 k_m^4 \qquad (6.47b)$$

which reduces to $(\text{Im}\omega_c)^2 = A^2$ at the instability threshold (k_{m_c}, B_c). We see, again, that $\text{Im}\omega_c$ is completely intrinsic, as it is determined entirely by the system's parameters. Notice that criticality at $k_{m_c} = 0$ is forbidden for fixed boundary conditions. In this case the critical mode will be the mode lying closest to zero allowed by these conditions.

Let us now turn to the possibility that the oscillatory instability is accompanied by space symmetry-breaking. To realize this, systems involving more than two variables are necessary. To see how this can happen we write the characteristic equation (6.39) of a three-variable system in the form

$$\omega^3 - T(k_m)\omega^2 + \delta(k_m)\omega - \Delta(k_m) = 0 \qquad (6.48)$$

where T and Δ have the same interpretation as in eq. (4.17a) and δ is the sum of principal minors of rank two of the coefficient matrix of eqs. (6.38) with $\omega = 0$. Let us look at the conditions imposed on (6.48) by an oscillatory instability. We have, at the instability threshold,

$$\omega_{1,2} = \pm i\Omega$$

$$\omega_3 < 0$$

Inserting into eq. (6.48) and equating real and imaginary parts we find:

$$T(\lambda, k_m) = \omega_3 < 0$$

$$\delta(\lambda, k_m) = \Omega^2 > 0$$

$$\Delta(\lambda, k_m) = \Omega^2 \omega_3 < 0$$

from which follows a relation linking λ to k_m,

$$T(\lambda, k_m)\delta(\lambda, k_m) = \Delta(\lambda, k_m) \qquad (6.49)$$

The point is that this equation is of sixth degree in k_m. As a result the λ versus k_m curve can have a minimum λ_c at a nontrivial value k_{m_c}, which is precisely the property that we wanted to establish.

C Competing instabilities

From the above discussion it follows that a given system can exhibit several different kinds of instability associated with Fig. 6.4(*a*) or (*b*), the imaginary part of the critical eigenvalue at the threshold being possibly zero or nonzero, depending on the system under consideration. In a concrete physical situation the particular type of instability that will be realized first will be the one associated with the lowest lying threshold, as λ is gradually increased from values corresponding to asymptotic stability in Fig. 6.3. This, in turn, is determined by the parameters present in the problem other than the one controlling stability. As an example let us derive the condition under which the Turing instability precedes the oscillatory instability in the Brusselator. Referring to Fig. 6.3 we must require that the minimum value B_c given by (6.45a) is less than the value given by (6.47a) evaluated at $k_m = 0$,

$$\left[1 + \left(\frac{D_1}{D_2} \right)^{1/2} A \right]^2 < A^2 + 1$$

This imposes the following inequality on the ratio of diffusion coefficients,

$$\left(\frac{D_1}{D_2} \right)^{1/2} < \frac{-1 + (A^2 + 1)^{1/2}}{A} < 1 \tag{6.50}$$

entailing that the autocatalytic species X must diffuse less efficiently than species Y. This could be anticipated from the intuitive argument that in the case of fast diffusion of the autocatalytic species (responsible for the instability) the nascent spatial pattern would tend to invade the whole system, thereby reestablishing the initial uniformity.

When the equality sign is realized in (6.50) the thresholds for Turing and oscillatory instabilities will coincide. This is a higher codimension instability near which the complex phenomena already alluded to in Section 5.8 could arise.

Another mechanism by which criticalities associated with Fig. 6.3 could compete is when, for a given value of the control parameter λ, the marginal stability curve for a given type of instability, say curve(*b*), has two extrema at two different values of k_m. Since the $\lambda(k_m^2)$ curve is symmetric with respect to the λ-axis we need for this at least four variables. In such a situation, which in the language of Section 5.8 corresponds also to a higher codimension instability, the linearized operator would have two nontrivial zero eigenfunctions at criticality (see eq. (6.36)). One would then be led to a situation similar to the one analyzed in eqs. (5.65)–(5.68). Typically the two values of k_m will be incommensurate. This can lead to complex, spatially quasi-periodic patterns arising from the mixing of the corresponding modes.

In a small system the above two-mode competition can be realized in an alternative way. Indeed, remembering that only the values $(2\pi/L)m$, m integer, are permitted for k_m, we may require that at a given value of λ slightly above the (generally inaccessible) minimum λ_c of curve(b) in Fig. 6.3 two permitted values of k_m lie on the marginal stability curve. This is again a higher codimension instability which can be handled by the methods of Section 5.8 (Mahar and Matkowsky, 1977).

6.5 Further comments on linear stability in spatially distributed systems

The analysis of Sections 6.3 and 6.4 suggests strongly that the transition to instability in spatially distributed systems presents some universal properties that are largely independent of the details of the particular system under consideration. All that seems to matter is:

the form of marginal stability curve $\lambda = \lambda(k_m)$, which depends essentially on the number of relevant variables present;
the type of instability involved ($\omega_c = 0$ or $\mathrm{Re}\,\omega_c = 0$, $\mathrm{Im}\,\omega_c \neq 0$ at criticality).

Furthermore, if the reference state displays appropriate symmetries the marginal stability curve depends on k_m^2 only, and the system properties are invariant under the transformation $k \rightarrow -k$. It is therefore not surprising that instabilities and bifurcations similar to those observed in the Bénard problem and in reaction–diffusion systems occur in a great variety of other systems as well, such as the Taylor vortex flow (Koschmieder, 1981), dynamic solidification (Langer, (1980), laser physics (Newell and Maloney, 1992; Lugiato, 1992; Lugiato and Lefever, 1987) and so on.

A distinct feature of instabilities in spatially distributed systems relates to the type and to the number of modes that can be destabilized beyond the threshold. Consider the Bénard problem. According to eq. (6.24) the solutions depend on two wave numbers k_x, k_y in the horizontal plane. On the other hand linear stability only fixes to the norm k_c of the vector $\mathbf{k} = (k_x, k_y)$ at the threshold (eq. (6.31)),

$$k_x^2 + k_y^2 = k_c^2 = \pi^2/2 \qquad (6.51a)$$

Clearly, this unique relation may be satisfied by various combinations of

k_x and k_y. In particular the ratio k_x/k_y, defining the orientation of the vector **k** in the horizontal plane, can be arbitrary. We conclude that the eigenvalue $\omega_c = 0$ of the linearized operator is *infinitely degenerate* at the instability threshold, since there is an infinity of eigenfunctions compatible with the marginal stability condition.

In view of eq. (6.24) it is clear that different choices of (k_x, k_y) will generate, at least at the level of linear analysis, different spatial patterns. As an example, the choice $k_x = k_c$, $k_y = 0$ gives rise at criticality to a vertical component of the velocity field equal to

$$w = w_k \sin \pi z \cos k_c x + cc$$

or, setting $w_k = (r_k/2) e^{i\phi}$, ϕ being a phase factor and $r_k/2$ a (positive) amplitude.

$$w = r_k \sin \pi z \cos (k_c x + \phi) \tag{6.51b}$$

This corresponds to an ascending maximal flux at $x = (1/k_c)(2n\pi - \phi)$, n integer, and to a descending maximal flux at $x = (1/k_c)[(2n + 1)\pi - \phi]$, n integer, representing a regular succession of rolls parallel to the y-axis. Similarly, the choice $k_x = k_y = k_c/\sqrt{2}$ gives rise at criticality to a vertical component w of the form (up to a phase factor)

$$w = r_k \sin \pi z \cos \left(\frac{k_c}{\sqrt{2}} x\right) \cos \left(\frac{k_c}{\sqrt{2}} y\right) \tag{6.51c}$$

This corresponds to a situation in which the fluid moves upwards near the center of a square and downwards along its sides, representing a regular succession of squares in the (x, y) plane.

An additional complication, to which we have already alluded in Section 6.4, appears when the critical threshold λ_c is exceeded. In a small system, since $k_x, \ldots$ are 'quantized', in the sense $k_x = (2\pi/L_x)n_x$, n_x integer (for periodic boundary conditions), there will always be a value λ close to λ_c such that only one mode corresponding to a particular choice of $n_x, \ldots$ will become unstable. But in a large system $(L_x \to \infty)$ the spectrum of k values becomes continuous. For any supercritical λ as close to λ_c as desired there will, then, inevitably be a whole continuum (k_1, k_2) of k values (Fig. 6.5) corresponding to positve eigenvalues of the linearized operator. We are thus confronted with a complex problem of *selection* of the solutions. This problem is far from being academic: it is manifested forcefully in the results of many of the experiments surveyed in Sections 1.3 and 1.4.

In view of the above comments one is entitled to express reservations about the very applicability of the techniques of bifurcation analysis to

Fig. 6.5 The zone of
unstable modes
(shaded area) in the
vicinity of the ($\omega_c = 0$,
$k_{m_c} \neq 0$) instability.

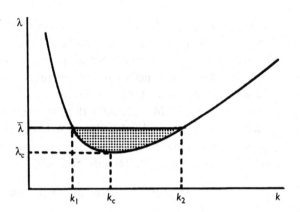

spatially distributed systems since, for one thing, in the presence of infinite degeneracy none of the known theorems (Section 5.1) guaranteeing bifurcation seem to apply. In the next sections we shall discard this difficulty and derive amplitude equations for the bifurcating solutions using perturbative expansions. We close the present section by stressing that in the light of some unexpected recent results (Constantin, Foias, Nicolaenko and Temam, 1989), this is most probably not a futile exercise. Specifically, for certain types of evolution equations of spatially distributed systems, one can show that the trajectories tend to a *finite-dimensional* (generally fractal) attractor embedded in a finite-dimensional phase space, referred to as the *inertial manifold*. This result may be impractical in the sense that these dimensionalities may be quite high in real-world problems. Still, it is of fundamental importance as it shows that the complications arising from the presence of an infinity of degrees of freedom may to a certain extent be bypassed.

6.6 Bifurcation analysis: general formulation

We now turn to the construction of the bifurcating solutions of the full nonlinear equations (6.1) beyond the instability threshold. As in eq. (5.11) we express these equations in terms of the excess variable, x_i around the reference state X_{is},

$$\partial \mathbf{x}(\mathbf{r}, t)/\partial t = \mathscr{L}(\lambda, \nabla) \cdot \mathbf{x}(\mathbf{r}, t) + \mathbf{h}(\lambda, \nabla, \mathbf{x}(\mathbf{r}, t)) \qquad (6.52)$$

where both the linearized operator $\mathscr{L}$ and the nonlinear contribution $\mathbf{h}$ contain now space derivatives acting on $\mathbf{x}(\mathbf{r}, t)$. We assume that linear stability analysis performed along the lines of Section 6.3 and 6.4 has revealed the existence of one of the above discussed criticalities (Figs. 6.1–6.4), entailing that at a certain value λ_c of the control parameter the linearized equation admits solutions of the form

$$\mathbf{x} = \mathbf{u}\, e^{i(\mathrm{Im}\,\omega_c)t} \phi_m(\mathbf{r})$$

$$= \mathbf{u}\, e^{i\Omega_c t} \phi_m(\mathbf{r}) \tag{6.53}$$

where $\phi_m(\mathbf{r})$ is an eigenfunction of the space-dependent part of $\mathscr{L}$ associated with the wave number k_m and satisfying the appropriate boundary conditions.

To compute the solution of (6.52) when λ deviates slightly from λ_c we expand, as in Section 5.2, both $\mathbf{x}$ and $\lambda - \lambda_c$ in power series of a smallness parameter ε, excluding for the moment discontinuities arising for instance from limit point bifurcations:

$$\mathbf{x} = \varepsilon \mathbf{x}_1 + \varepsilon^2 \mathbf{x}_2 + \cdots \tag{6.54}$$

$$\lambda - \lambda_c = \varepsilon \gamma_1 + \varepsilon^2 \gamma_2 + \cdots \tag{6.55}$$

Furthermore, to account for critical slowing down we introduce slow time scales according to

$$\frac{\partial}{\partial t} = \varepsilon \frac{\partial}{\partial \tau_1} + \varepsilon^2 \frac{\partial}{\partial \tau_2} + \cdots \tag{6.56}$$

if $\omega_c = 0$ at criticality, or

$$\frac{\partial}{\partial t} = \Omega_c \frac{\partial}{\partial T} + \varepsilon \frac{\partial}{\partial \tau_1} + \varepsilon^2 \frac{\partial}{\partial \tau_2} + \cdots \tag{6.57}$$

if $\mathrm{Re}\,\omega_c = 0$, $\mathrm{Im}\,\omega_c \neq 0$ at criticality.

It remains now to see how to handle spatial derivatives in (6.52). Two cases need to be distinguished.

A Systems of small spatial extent

As discussed in the previous sections the spectrum of k_m is discrete in such systems. For λ sufficiently close to λ_c only one (or a small number of) mode(s) will be destabilized. The corresponding $\phi_m(\mathbf{r})$ in (6.53) will then determine the space dependence of the dominant part of the solution in the nonlinear range as well. In this case the action of space derivatives in (6.52) is straightforward and no additional scaling needs to be performed.

B Systems of large spatial extent

As soon as one enters the unstable domain a continuum of modes will be excited. Owing to the nonlinear term in (6.52) these modes will interact. If one therefore formally performs an eigenfunction expansion of the type of eq. (3.2) for **x** one will find that the equations for the expansion coefficients (eq. (3.3)) will involve coefficients associated with linear combinations of the k_ms of the different modes. Expressed in terms of the variable **r** rather than k_m, this entails that the system involves more than one space scale.

In order now to infer the dominant scales among the continuum of the scales present we express k_m as its reference value k_{m_c} (the latter being zero, Fig. 6.4(a), or finite, Fig. 6.4(b)) plus a deviation Δk,

$$k_m = k_{m_c} + \Delta k \tag{6.58}$$

and observe that at a given distance from the instability threshold $\lambda - \lambda_c$, the range of Δk can be estimated by (cf. eq. (6.42))

$$\Delta k \approx \left[\frac{a}{b}(\lambda - \lambda_c)\right]^{1/2} \tag{6.59}$$

Accordingly a space derivative acting on $\mathbf{x}(\mathbf{r}, t)$ will give rise to a 'fast' variation associated with k_{m_c} and a 'slow' one associated with Δk, the latter being weighted by a factor $(\lambda - \lambda_c)^{1/2}$ or, in the dominant order, by a factor $\varepsilon^{1/2}$ or ε according as $\gamma_1 \neq 0$ or $\gamma_1 = 0$ in (6.55):

$$\frac{\partial}{\partial r} = \frac{\partial}{\partial R} + \varepsilon^{1/2}\frac{\partial}{\partial \rho_1} + \cdots (\gamma_1 \neq 0) \tag{6.60a}$$

$$\frac{\partial}{\partial r} = \frac{\partial}{\partial R} + \varepsilon\frac{\partial}{\partial \rho_1} + \cdots (\gamma_1 = 0) \tag{6.60b}$$

Substituting (6.54)–(6.57) and (6.60) into eqs. (6.52) we get, to the various orders in ε, a set of linear equations for the successive approximations $\mathbf{x}_k$. The $O(\varepsilon)$ equations are homogeneous and have the same structure as the linear stability equations. Their solution is, therefore, of the form (cf. (6.53))

$$\mathbf{x}_1(\mathbf{r}, t) = c(\rho, \tau)\mathbf{u}\,e^{i\Omega_c T}\phi_m(\mathbf{R}) + \text{cc} \tag{6.61}$$

where we have introduced, as in eq. (5.13), an amplitude c. This amplitude is undetermined at this stage and is allowed to depend on the slow time *and* space scales that do not appear explicitly in the $O(\varepsilon)$ equations. It plays in this respect the role of an *envelope* modulating the variation expressed by $\phi_m(\mathbf{R})$.

To the next orders the equations for $\mathbf{x}_k$, $k \geq 2$ are inhomogeneous. To compute them one has to impose, in the spirit of Chapter 5, *solvability conditions* since the operator $\mathscr{L}(\lambda_c, \nabla)$ acting on $\mathbf{x}_k$ has a nontrivial null space. These conditions provide one with equations for the so far undetermined amplitude c, which are the analogs of the normal form equations of Chapter 5. The main novelty with respect to Chapter 5 is that in systems of large spatial extent these equations are going to be *partial differential equations*, owing to the dependence of c on the slow space variable ρ. This procedure is illustrated in the next few sections on a number of representative examples considering, to begin with, the case of small systems.

6.7 Bifurcation of two-dimensional rolls in the Bénard problem: the small aspect ratio case

Our first illustration of the machinery of bifurcation theory in spatially distributed systems will be the, by now familiar, Bénard problem in the Boussinesq approximation and with free boundary conditions. The starting point is the nonlinear Boussinesq equations derived in Section 2.5. As shown in Section 6.2 there is a definite advantage in expressing the evolution laws of physical systems in dimensionless form. We therefore apply to these equations the scaling used in the study of linear stability analysis (eqs. (6.11)). A straightforward algebra leads to

$$
\left.
\begin{aligned}
&\partial\theta/\partial t + (\delta\mathbf{v}\cdot\nabla)\theta = Rw + \nabla^2\theta \\[4pt]
&\partial\delta\mathbf{v}/\partial t + (\delta\mathbf{v}\cdot\nabla)\delta\mathbf{v} = P(-\nabla\delta p + \nabla^2\delta\mathbf{v} + \theta\mathbf{1}_z) \\[4pt]
&\operatorname{div}\delta\mathbf{v} = 0
\end{aligned}
\right\} \quad (6.62)
$$

where the dimensionless parameters R and P have been defined in eq. (6.13). To simplify as much as possible we shall assume that the Prandtl number is very high, $P \gg 1$. Dividing through the second equation (6.62) by P we see that one can neglect, in this limit, the nonlinearity $(\delta\mathbf{v}\cdot\nabla)\delta\mathbf{v}$ compared to $(\delta\mathbf{v}\cdot\nabla)\theta$ as well as the contribution of the acceleration term $\partial\delta\mathbf{v}/\partial t$. Furthermore, to eliminate the pressure term from this equation we apply the curl operator twice and project on the z-axis. We finally obtain, following (6.17), the following simplified set of equations

$$
\left.
\begin{aligned}
&\partial\theta/\partial t + (\delta\mathbf{v}\cdot\nabla)\theta = Rw + \nabla^2\theta \\[4pt]
&\frac{\partial^2\theta}{\partial x^2} + \frac{\partial^2\theta}{\partial y^2} + \nabla^4 w = 0 \\[4pt]
&\operatorname{div}\delta\mathbf{v} = 0
\end{aligned}
\right\} \quad (6.63)
$$

subject to the boundary conditions (6.22).

As stressed throughout Chapter 1, the evolution laws of a spatially distributed system can generate a great variety of patterns. One of the principal reasons for this diversity, noted in Section 6.5, is the degeneracy in the orientation of the wave vector $\mathbf{k}$. In what follows we shall seek specific types of solution which for simplicity we take here in the form of rolls. According to eq. (6.51b) this type of structure is two-dimensional. At the level of eq. (6.63) this implies

$$\left.\begin{array}{l} \delta\mathbf{v} = u(x, z)\mathbf{1}_x + w(x, z)\mathbf{1}_z \\[2mm] \theta = \theta(x, z) \end{array}\right\} \quad (6.64)$$

We are now in the position to perform the bifurcation analysis of the solutions. We expand θ, $\delta\mathbf{v}$, R and $\partial/\partial t$ according to (6.54)–(6.56),

$$\left.\begin{array}{l} \theta = \varepsilon\theta_1 + \varepsilon^2\theta_2 + \cdots \\[2mm] \delta\mathbf{v} = \varepsilon\delta\mathbf{v}_1 + \varepsilon^2\delta\mathbf{v}_2 + \cdots \\[2mm] R = R_c + \varepsilon R_1 + \cdots \\[2mm] \dfrac{\partial}{\partial t} = \varepsilon\dfrac{\partial}{\partial\tau_1} + \varepsilon^2\dfrac{\partial}{\partial\tau_2} + \cdots \end{array}\right\} \quad (6.65)$$

and substitute into eq. (6.63). We keep spatial coordinates unscaled by limiting ourselves to a small aspect ratio. To $O(\varepsilon)$ we obtain, taking (6.64) into account,

$$\left.\begin{array}{l} \nabla^2\theta_1 + R_c w_1 = 0 \\[2mm] \partial^2\theta_1/\partial x^2 + \nabla^4 w_1 = 0 \end{array}\right\} \quad (6.66)$$

which is nothing but the linearized problem of Section 6.3 evaluated at criticality. Its solution is, therefore, of the form (cf. eqs. (6.20), (6.24) and (6.61)),

$$\binom{\theta_1}{w_1} = \binom{1}{\psi}\sin\pi z[c(\tau_1\ldots)\,\mathrm{e}^{\mathrm{i}k_c x} + c^*(\tau_1\ldots)\,\mathrm{e}^{-\mathrm{i}k_c x}] \quad (6.67)$$

In accordance with the formulation of Section 6.6 we have introduced a (complex) amplitude c, which remains undetermined at this stage. Accordingly we have normalized the Fourier coefficients θ_{kn}, w_{kn} in such a way that $\theta_{kn} = 1$ and $\psi = w_{kn}/\theta_{kn}$. The latter is computed from the linearized equation (6.25) at criticality:

$$\psi = \frac{k_c^2 + \pi^2}{R_c} \quad (6.68)$$

Since the flow is two-dimensional (eq. (6.64)) we also need, in this order, information on the horizontal component $u(x, z)$. The most straightforward way to get this is to use the incompressibility condition (the third relation of (6.63))

$$\partial u_1/\partial x = -\partial w_1/\partial z$$

and observe that for periodic rolls it is satisfied by the expression

$$u_1 = \frac{1}{k_c^2}\frac{\partial^2 w_1}{\partial x \partial z} \qquad (6.69\text{a})$$

Indeed, differentiating (6.69a) with respect to x and realizing from (6.51b) that the second derivative of w_1 with respect to this variable reproduces w_1 up to a factor $(-k_c^2)$, one obtains the incompressibility condition as an identity. Substituting w_1 from (6.67) one obtains the more explicit form

$$u_1 = \frac{i\pi\psi}{k_c}\cos \pi z(c\, e^{ik_c x} - c^*\, e^{-ik_c x}) \qquad (6.69\text{b})$$

To compute the amplitude c in eq. (6.67) we have to go to the higher orders in the perturbation expansion. We discuss first the $O(\varepsilon^2)$ and then the $O(\varepsilon^3)$ equations.

A The $O(\varepsilon^2)$ equations

Eqs. (6.63)–(6.65) lead to the following set of equations for the unknowns (θ_2, w_2):

$$\mathscr{L}(R_c,\nabla)\begin{pmatrix}\theta_2\\w_2\end{pmatrix} = \begin{pmatrix}\nabla^2 & R_c\\ \dfrac{\partial^2}{\partial x^2} & \nabla^4\end{pmatrix}\begin{pmatrix}\theta_2\\w_2\end{pmatrix}$$

$$= \begin{pmatrix}\dfrac{\partial\theta_1}{\partial\tau_1} - R_1 w_1 + u_1\dfrac{\partial\theta_1}{\partial x} + w_1\dfrac{\partial\theta_1}{\partial z}\\ 0\end{pmatrix} = \begin{pmatrix}q_2\\0\end{pmatrix} \qquad (6.70)$$

The right hand side can be evaluated using (6.67) and (6.69b). One obtains

$$\begin{pmatrix}q_2\\0\end{pmatrix} = \begin{pmatrix}\sin \pi z\left(\dfrac{\partial c}{\partial\tau_1}e^{ik_c x} + cc\right) - R_1\psi \sin \pi z(c\, e^{ik_c x} + cc) + 2\pi\psi\,|c|^2\sin 2\pi z\\ 0\end{pmatrix}$$

$$(6.71)$$

The solvability of (6.70) requires that $\begin{pmatrix}q_2\\0\end{pmatrix}$ be orthogonal to the null space of $\mathscr{L}^+(R_c,\nabla)$. One checks easily that with the boundary conditions adopted

$$\mathscr{L}^+(R_c,\nabla) = \begin{pmatrix}\nabla^2 & \dfrac{\partial^2}{\partial x^2}\\ R_c & \nabla^4\end{pmatrix} \qquad (6.72\text{a})$$

and admits, for a zero eigenvalue, eigenfunctions with the same spatial dependence as those of $\mathscr{L}(R_c, V)$,

$$\mathbf{u}^+ = \begin{pmatrix} 1 \\ \psi^+ \end{pmatrix} e^{ik_c x} \sin \pi z \qquad (6.72b)$$

where ψ^+ is computed from $\mathscr{L}^+ \mathbf{u}^+ = 0$ at criticality.

We may now express the solvability condition explicitly, adopting a Hilbert space scalar product in addition to the usual one of vector calculus, to account for the space dependence of the functions involved:

$$\left(\begin{pmatrix} 1 \\ \psi^+ \end{pmatrix} e^{ik_c x} \sin \pi z, \begin{pmatrix} q_2 \\ 0 \end{pmatrix} \right) = 0 \qquad (6.73)$$

or, more explicitly, using the fact that the functions under the scalar product are periodic in x,

$$\int_0^1 dz \int_0^{\frac{2\pi}{k_c}} dx \left[\sin^2 \pi z \left(\frac{\partial c}{\partial \tau_1} + \frac{\partial c^*}{\partial \tau_1} e^{-2ik_c x} \right) - R_1 \psi \sin^2 \pi z (c + c^* e^{-2ik_c x}) \right.$$
$$\left. + 2\pi \psi |c|^2 \sin \pi z \sin 2\pi z \, e^{-ik_c x} \right] = 0$$

Using the property that the integral of $e^{imk_c x}$ over a period $2\pi/k_c$ vanishes for $m \neq 0$ one sees that the above equation reduces to

$$\partial c/\partial \tau_1 = R_1 \psi c$$

As pointed out in Section 5.4, this linear equation predicts unphysical runaway behavior beyond bifurcation. The only way to avoid this deficiency is to reduce it to a trivial identity, which can be achieved by setting

$$\left. \begin{aligned} R_1 &= 0 \\ \partial c/\partial \tau_1 &= 0 \end{aligned} \right\} \qquad (6.74)$$

The solvability condition (6.73) being thus satisfied one may solve the simplified second order equations (6.70):

$$\nabla^2 \theta_2 + R_c w_2 = 2\pi \psi |c|^2 \sin 2\pi z$$
$$\partial^2 \theta_2/\partial x^2 + \nabla^4 w_2 = 0 \qquad (6.75)$$

To this end we eliminate w_2 by applying the operator ∇^4 to the first equation and subtracting the second equation multiplied by R_c. This yields:

$$\left(\nabla^6 - R_c \frac{\partial^2}{\partial x^2} \right) \theta_2 = (2\pi)^5 \psi |c|^2 \sin 2\pi z \qquad (6.76)$$

Expanding θ_2 in Fourier series one arrives at

$$\theta_2 = -\frac{1}{2\pi}\psi|c|^2 \sin 2\pi z \tag{6.77a}$$

Substituting into the second of eqs. (6.75) and taking into account the free boundary conditions, eq. (6.22), one arrives at the trivial solution

$$w_2 = 0 \tag{6.77b}$$

and hence, by virtue of (6.69a),

$$u_2 = 0 \tag{6.77c}$$

This completes the solution to the second order of perturbation theory.

B The $O(\varepsilon^3)$ equations

To the next order one is led, using eqs. (6.63)–(6.65), to the following set of equations for the unknowns (θ_3, w_3):

$$\mathscr{L}(R_c, \nabla)\begin{pmatrix} \theta_3 \\ w_3 \end{pmatrix} = \begin{pmatrix} \nabla^2 & R_c \\ \dfrac{\partial^2}{\partial x^2} & \nabla^4 \end{pmatrix} \begin{pmatrix} \theta_3 \\ w_3 \end{pmatrix}$$

$$= \begin{pmatrix} \dfrac{\partial \theta_1}{\partial \tau_2} - R_2 w_1 + u_1 \dfrac{\partial \theta_2}{\partial x} + w_1 \dfrac{\partial \theta_2}{\partial z} + u_2 \dfrac{\partial \theta_1}{\partial x} + w_2 \dfrac{\partial \theta_1}{\partial z} \\ 0 \end{pmatrix}$$

$$= \begin{pmatrix} q_3 \\ 0 \end{pmatrix} \tag{6.78}$$

The right hand side can be evaluated explicitly using expressions (6.67), (6.69b) and (6.77a)–(6.77c). One finds

$$\begin{pmatrix} q_3 \\ 0 \end{pmatrix} = \begin{pmatrix} \sin \pi z\left(\dfrac{\partial c}{\partial \tau_2}e^{ik_c x} + \mathrm{cc}\right) + \psi \sin \pi z(c\, e^{ik_c x} + \mathrm{cc})(-R_2 - \psi|c|^2 \cos 2\pi z) \\ 0 \end{pmatrix} \tag{6.79}$$

The solvability of (6.78) now requires $\begin{pmatrix} q_3 \\ 0 \end{pmatrix}$ to be orthogonal to the null eigenvector of $\mathscr{L}^+$ introduced earlier in this section. Utilizing the same definition of the Hilbert space scalar product along with the periodicity of the solutions in the x-direction we obtain

$$\int_0^1 dz \int_0^{\frac{2\pi}{k_c}} dx\, e^{-ik_c x} \sin^2 \pi z \left[\frac{\partial c}{\partial \tau_2}e^{ik_c x} + \mathrm{cc}\right.$$

$$\left. + \psi(c\, e^{ik_c x} + \mathrm{cc})(-R_2 - \psi|c|^2 \cos 2\pi z)\right] = 0 \tag{6.80}$$

After explicit evaluation of the integrals one ends up with the equation

$$\frac{\partial c}{\partial \tau_2} = R_2 \psi c - \tfrac{1}{2}\psi^2 |c|^2 c$$

As in Section 5.4 we introduce the normalized amplitude $z = \varepsilon c$ and reestablish from (6.65) the initial parameters t and $R - R_c$. This leads to

$$\partial z/\partial t = (R - R_c)\psi z - \tfrac{1}{2}\psi^2 |z|^2 z \qquad (6.81)$$

which has the same structure as the normal form of a pitchfork bifurcation. The Bénard problem in a system of small aspect ratio close to the bifurcation of convective solutions in the form of rolls has thus been reduced to a single ordinary differential equation for the amplitude of the solution to the dominant order. The latter admits the nontrivial steady state value

$$|z| = \left[\frac{2}{\psi}(R - R_c)\right]^{1/2} \qquad (6.82)$$

which is nonanalytic in the parameter $R - R_c$. It bifurcates supercritically and is, therefore, stable. Notice that the phase of the solution cannot be determined by the third order solvability condition nor, in fact, by higher order perturbation analysis. This was to be expected since, as noticed in Section 1.3, in the absence of lateral boundaries there is no mechanism capable of fixing *a priori* the position of a convection cell along the x-axis.

The solution of the Bénard problem to the dominant order can now be written explicitly. Using eqs. (6.67) and (6.82) we obtain

$$\begin{pmatrix} \theta(x, z) \\ w(x, z) \end{pmatrix} = \varepsilon \begin{pmatrix} \theta_1 \\ w_1 \end{pmatrix} = \left(\frac{2}{\psi}\right)^{1/2} (R - R_c)^{1/2} \begin{pmatrix} 1 \\ \psi \end{pmatrix} e^{i(k_c x + \phi)} \sin \pi z + \mathrm{cc}$$
$$+ O(R - R_c) \qquad (6.83)$$

where ϕ denotes the phase. We have thus established analytically the phenomenon of *space symmetry-breaking* which, as mentioned earlier in this chapter, is one of the principal signatures of the complexity of nonlinear spatially distributed systems under nonequilibrium constraint.

6.8 Bifurcation analysis in systems of large spatial extent: complex Landau–Ginzburg equation

We now turn to systems whose extent in at least one direction is much larger that the characteristic length of the critical mode predicted by linear stability analysis. As pointed out in Section 6.6B, the bifurcation analysis for such systems has to be completed by an appropriate scaling of spatial coordinates as well, (6.60a)–(6.60b). We already know that the specific type of scaling depends on the type of instability experienced. In the present section we focus on a reaction – diffusion type system in the vicinity of an instability corresponding to a pair of complex eigenvalues crossing the imaginary axis at criticality, Section 6.4B. To be even more specific we suppose that in the system under consideration the first instability occurs at $k_{m_c} = 0$, a situation which we referred to in Section 6.4 as time symmetry-breaking. The dispersion relation for the real part of the eigenvalue will then be as in eq. (6.42a) and Fig. 6.4(*a*),

$$\text{Re } \omega \approx a(\lambda - \lambda_c) - bk_m^2 + O(k_m^4) \tag{6.84a}$$

while the imaginary part of the eigenvalue will be given by an equation of the type of (6.47b),

$$\text{Im } \omega \approx \Omega_c + c(\lambda - \lambda_c) + gk_m^2 + O(k_m^4) \tag{6.84b}$$

Accordingly, one can anticipate a scaling of the time variable as in eq. (6.57a). The scaling of the space variable will be similar to eq. (6.60), but with the 'fast' variation $\partial/\partial R$ absent since the most unstable (and hence dominant) modes will display close to instability ($\text{Re}\omega = 0$) space dependences modulated by small wave numbers. Recalling from the analysis of the Hopf bifurcation in spatially uniform systems (Section 5.7) that $\gamma_1 = 0$ and $\Omega_1 = 0$ in eqs. (6.55) and (6.57b), owing to the periodicity in the time variable, one is finally led to the perturbative analysis of eqs. (6.52),

$$\partial \mathbf{x}/\partial t = \mathscr{L}(\lambda, \nabla) \cdot \mathbf{x} + \mathbf{h}(\lambda, \nabla, \mathbf{x})$$

with

$$\mathbf{x} = \varepsilon \mathbf{x}_1 + \varepsilon^2 \mathbf{x}_2 + \cdots \tag{6.85a}$$

$$\lambda - \lambda_c = \varepsilon^2 \gamma_2 + \cdots \tag{6.85b}$$

$$\frac{\partial}{\partial t} = \Omega_c \frac{\partial}{\partial T} + \varepsilon^2 \frac{\partial}{\partial \tau} + \cdots \tag{6.85c}$$

$$\frac{\partial}{\partial r} = \varepsilon \frac{\partial}{\partial \rho} + \cdots \tag{6.85d}$$

Actually, in a reaction–diffusion type system with constant diffusion coefficients one deals with a more restricted form of eq. (6.52), whereby $\mathbf{h}$ is ∇-independent and $\mathscr{L}$ can be split into a ∇-independent part plus a contribution proportional to the Laplace operator:

$$\partial \mathbf{x}/\partial t = \mathscr{L}_0(\lambda) \cdot \mathbf{x} + \mathscr{L}_1(\lambda) \nabla^2 \mathbf{x} + \mathbf{h}(\mathbf{x}, \lambda) \tag{6.86}$$

More generally, in eq. (6.52) $\mathscr{L}$ and $\mathbf{h}$ can be expanded formally in series containing increasingly high derivatives in space. The structure of this series depends critically on symmetry properties – for instance, in an isotropic system only even derivatives will show up. Eq. (6.86) can therefore be regarded as the simplest realization of this structure.

Substituting now (6.85a)–(6.85d) into (6.86) and proceeding as in Sections 5.2, 5.7 and 6.7 we get, to the various orders in ε, the following systems of equations.

A $O(\varepsilon)$

$$\left(\Omega_c \frac{\partial}{\partial T} \mathbf{1} - \mathscr{L}_0(\lambda_0)\right) \cdot \mathbf{x}_1 = 0 \tag{6.87}$$

This homogeneous system of equations is equivalent to linear stability analysis. It therefore admits solutions of the form

$$\mathbf{x}_1 = c(\tau, \rho)\mathbf{u}\, e^{iT} + \text{cc} \tag{6.88}$$

B $O(\varepsilon^2)$

One obtains (cf. eq. (5.11)) the inhomogeneous system of equations

$$\left[\Omega_c \frac{\partial}{\partial T} \mathbf{1} - \mathscr{L}_0(\lambda_c)\right] \cdot \mathbf{x}_2 = \frac{1}{2}\mathbf{h}_{xx} \cdot \mathbf{x}_1 \mathbf{x}_1 \tag{6.89}$$

Having set at the very beginning $\gamma_1 = 0$, $\Omega_1 = 0$ the solvability condition is automatically satisfied. In view of (6.88) one sees that the right hand side contains contributions in e^{2iT} and e^{-2iT}, as well as terms independent of T. We may therefore seek solutions $\mathbf{x}_2$ of the form

$$\mathbf{x}_2 = c^2 \mathbf{p}_2\, e^{2iT} + c^{*2} \mathbf{p}_2^*\, e^{-2iT} + |c|^2 \mathbf{p}_0 \tag{6.90}$$

it being understood that the coefficients $\mathbf{p}_2$ and $\mathbf{p}_0$ can be determined uniquely once the detailed structure of the system is specified.

C $O(\varepsilon^3)$

To this order the slow space and time dependences are manifested for the first time. One obtains (cf. also (5.12)):

$$\left(\Omega_c \frac{\partial}{\partial T} \mathbf{1} - \mathscr{L}_0(\lambda_c)\right) \cdot \mathbf{x}_3 = \gamma_2 \mathscr{L}_{0\lambda}(\lambda_c) \cdot \mathbf{x}_1 + \mathbf{h}_{xx}(\lambda_c) \cdot \mathbf{x}_1 \mathbf{x}_2$$

$$+ \frac{1}{6}\mathbf{h}_{xxx} \cdot \mathbf{x}_1 \mathbf{x}_1 \mathbf{x}_1 - \frac{\partial \mathbf{x}_1}{\partial \tau} + \mathscr{L}_1(\lambda_c)\nabla_\rho^2 \mathbf{x}_1 = \mathbf{q}_3 \tag{6.91}$$

The solvability condition of this inhomogeneous system of equations requires that the right hand side be orthogonal to the null eigenspace of the adjoint of the operator acting on x_3 which, as discussed in Section 5.7, is of the form $\mathbf{u}^+ \mathrm{e}^{\mathrm{i}T}$. We therefore obtain, following the previously adopted extended definition of scalar product:

$$\int_0^{2\pi} \mathrm{d}T\, \mathrm{e}^{-\mathrm{i}T} \mathbf{u}^{+*} \cdot \mathbf{q}_3(c, T, \rho) = 0 \qquad (6.92)$$

where the dot indicates the ordinary scalar product in a linear vector space.

Substituting the detailed form of $\mathbf{x}_1$ and $\mathbf{x}_2$ from eqs. (6.88) and (6.90) we further obtain

$$\int_0^{2\pi} \mathrm{d}T\, \mathrm{e}^{-\mathrm{i}T} \mathbf{u}^{+*} \cdot \left\{ \gamma_2 c [\mathscr{L}_{0\lambda}(\lambda_c) \cdot \mathbf{u}]\, \mathrm{e}^{\mathrm{i}T} - \frac{\partial c}{\partial \tau} \mathbf{u}\, \mathrm{e}^{\mathrm{i}T} + \nabla_\rho^2 c [\mathscr{L}_1(\lambda_c) \cdot \mathbf{u}]\, \mathrm{e}^{\mathrm{i}T} \right.$$

$$+ \frac{1}{2} (\mathbf{h}_{\mathbf{xxx}} \cdot \mathbf{uuu}^*) |c|^2 c\, \mathrm{e}^{\mathrm{i}T} + [\mathbf{h}_{\mathbf{xx}}(\lambda_c) \cdot (\mathbf{p}_2 \mathbf{u}^* + \mathbf{p}_0 \mathbf{u})]\, \mathrm{e}^{\mathrm{i}T} |c|^2 c$$

$$+ \left. (\text{terms in } \mathrm{e}^{3\mathrm{i}T}, \mathrm{e}^{2\mathrm{i}T}, \mathrm{e}^{-\mathrm{i}T}, \mathrm{e}^{-2\mathrm{i}T}, \mathrm{e}^{-3\mathrm{i}T}) \right\} \qquad (6.93)$$

Clearly, owing to the integration in T only the terms in $\mathrm{e}^{\mathrm{i}T}$ will survive in the curly brackets in eq. (6.93). This leads to an equation for the amplitude c of the form

$$(\mathbf{u}^+ \cdot \mathbf{u})\, \partial c/\partial \tau = \gamma_2 P_1 c - P_3 |c|^2 c + Q_1 \nabla_\rho^2 c \qquad (6.94)$$

in which the coefficients, P_1, P_3 and Q_1 are uniquely determined from (6.93) and the structure of the underlying system. We notice that, in general, P_3 and Q_1 are complex-valued. Dividing through by $\mathbf{u}^+ \cdot \mathbf{u}$, introducing the normalized amplitude $z = \varepsilon c$ and reestablishing the initial parameters $\lambda - \lambda_c$, $\mathbf{r}$ and t through eqs. (6.85) we finally obtain

$$\partial z/\partial t = (\lambda - \lambda_c)z + (1 + \mathrm{i}\alpha)\nabla^2 z - (1 + \mathrm{i}\beta)|z|^2 z \qquad (6.95)$$

where we have performed a further (ε-independent) scaling to eliminate P_1 and the real parts of P_3 and Q_1, both assumed to be positive.

With a certain degree of caution with regard to mathematical rigor in line with the comments made in Section 6.5, one may consider eq. (6.95) as the *normal form* of a spatially extended dynamical system in the vicinity of an instability of the type ($\mathrm{Re}\,\omega_c = 0$, $\mathrm{Im}\,\omega_c \neq 0$, $k_{m_c} = 0$). This equation brings a correction to the normal form of a Hopf bifurcation obtained earlier (eq. (5.53)) by introducing a slowly varying *envelope* modulating the amplitude of the individual oscillators in different points in space. This

particular way of describing the coupling between spatially distributed
oscillators induced by diffusion has the great merit of universality, at least
in the vicinity of the bifurcation point. It was first studied systematically
by Newell (1974) and Kuramoto (1984) and, more recently, by Coullet
and coauthors (Coullet and Gil, 1988) who place special emphasis on the
role of the symmetries built into the system.

Eq. (6.95) is referred to in the literature as the *complex Landau–
Ginzburg equation*, since it generalizes the (real-valued) Landau–Ginzburg equation familiar from equilibrium critical phenomena. The latter
corrects, in turn, Landau's mean field theory of critical phenomena to
which we alluded in Section 5.6, by allowing for spatially inhomogeneous
fluctuations (Ma, 1976). In the opposite limit of purely imaginary
coefficients and $\lambda = \lambda_c$, eq. (6.95) reduces to the nonlinear Schrödinger
equation familiar from the study of waves in nondissipative systems
(Newell and Maloney, 1992). The fact that in the present context the
coefficients of the cubic and spatial derivative terms are complex is a
consequence of both the dissipative character of the dynamics and the
nonequilibrium constraints. It entails that contrary to its equilibrium
counterpart, this equation does not derive from a potential. This opens
the way to new, specifically nonequilibrium phenomena arising from the
loss of stability of the homogeneous limit cycle and ranging from
propagating wave fronts to spatio-temporal chaos and the generation of
defects. Two characteristic illustrations obtained from direct numerical
simulation of (6.95) are depicted in Fig. 6.6. The analogy with the
phenomena described in Section 1.4 in connection with chemical
instabilities and defects is quite striking and highlights the importance of
the complex Landau–Ginzburg equation in large classes of natural
phenomena.

To get a flavor of how this complexity can arise it is useful to sketch the

Fig. 6.6 Spatio-
temporal complexity
generated by the
complex
Landau–Ginzburg
equation (6.95) in a
two-dimensional
system: (*a*) spiral wave
obtained for
$\lambda - \lambda_c = 1, \alpha = 1,$
$\beta = -0.6,$
$L_x = L_y = 50; (b)$
solution in the
unstable region
$1 + \alpha\beta < 0:$
$\lambda - \lambda_c = 1,$
$\alpha = 2, \beta = -0.82,$
$L_x = L_y = 50$ (Lega,
1989).

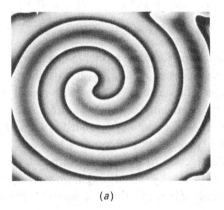
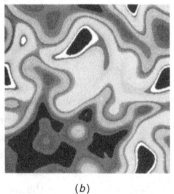

(*a*)   (*b*)

linear stability analysis of the homogeneous limit cycle solution of (6.95) in a one-dimensional system. From eqs. (5.55)–(5.57) we have

$$z_s = (\lambda - \lambda_c)^{1/2} \, e^{-i\beta(\lambda - \lambda_c)t} \tag{6.96}$$

Setting

$$z = z_s + \delta z(r, t) \, e^{-i\beta(\lambda - \lambda_c)t} \tag{6.97}$$

we obtain the following linearized version of (6.95):

$$\partial \delta z / \partial t = -(1 + i\beta)(\lambda - \lambda_c)(\delta z + \delta z^*) + (1 + i\alpha)\nabla^2 \delta z \tag{6.98}$$

or, with $\delta z = \overline{\delta z} \, e^{ikr}$, $\overline{\delta z} = u + iv$ and after separation of real and imaginary parts,

$$\left. \begin{array}{l} du/dt = -\,[2(\lambda - \lambda_c) + k^2]u + \alpha k^2 v \\[2mm] dv/dt = -\,[2\beta(\lambda - \lambda_c) + \alpha k^2]u - k^2 v \end{array} \right\} \tag{6.99}$$

The characteristic equation of this system reads (see Sections 4.3 and 6.4)

$$\omega^2 + 2[(\lambda - \lambda_c) + k^2]\omega + [2(\lambda - \lambda_c)k^2(1 + \alpha\beta) + k^4(1 + \alpha^2)] = 0 \tag{6.100}$$

In the supercritical region $\lambda - \lambda_c \geq 0$ the sum of the roots obviously remains negative. The only instability that can arise is, therefore, through the constant term vanishing and subsequently becoming negative. This requires that the parameters α and β be such that

$$1 + \alpha\beta < 0 \tag{6.101}$$

This inequality constitutes a condition for spatio-temporal complexity since in this range a homogeneous oscillation encompassing the system as a whole may no longer be sustained. We notice that the instability is induced by the presence of complex coefficients in eq. (6.95) or, equivalently, by the presence of a nontrivial *phase variable* related to the imaginary part of the order parameter z. In this sense, therefore, it can be referred to as a *phase instability*.

6.9 Further examples of normal form envelope equations in large systems

As stressed in Section 6.8 and throughout Chapter 5, the structure of normal form equations depends on the type of instability experienced by the system. In this section we briefly summarize the procedure leading to such equations in large systems in the vicinity of instabilities other than the (Re $\omega_c = 0$, Im $\omega_c \neq 0$, $k_{m_c} = 0$) one considered in the preceding section.

A The (Re $\omega_c = 0$, Im $\omega_c = 0$, $k_{m_c} \neq 0$) instability

This situation, which we referred to in Sections 6.3 and 6.4 as space symmetry-breaking, includes as particular cases the Bénard and Turing instabilities. The dispersion relation for the real part of the eigenvalue will be as in eq. (6.42b), entailing that the scaling of the time and space variables will be as in eqs. (6.56) and (6.60) respectively. In particular, one will have to account both for a 'fast' and for a 'slow' space variable, R and ρ.

Consider first a one-dimensional system. We anticipate that, because of spatial symmetries, $\gamma_1 = 0$ and $(\partial/\partial\tau_1)(\ldots) = 0$ in the expansion of eqs. (6.55) and (6.56). We are therefore led to solve eqs. (6.86) with the following perturbation scheme:

$$
\left.
\begin{aligned}
\mathbf{x} &= \varepsilon\mathbf{x}_1 + \varepsilon^2\mathbf{x}_2 + \cdots \\[4pt]
\lambda - \lambda_c &= \varepsilon^2\gamma_2 + \cdots \\[4pt]
\frac{\partial}{\partial t} &= \varepsilon^2\frac{\partial}{\partial\tau} \\[4pt]
\frac{\partial}{\partial\tau} &= \frac{\partial}{\partial R} + \varepsilon\frac{\partial}{\partial\rho}
\end{aligned}
\right\}
\qquad (6.102)
$$

Inserting in (6.86) we find, to the order $O(\varepsilon)$,

$$
\left[\mathscr{L}_0(\lambda_c) + \mathscr{L}_1(\lambda_c)\frac{\partial^2}{\partial R^2} \right]\cdot\mathbf{x}_1 = 0
\qquad (6.103)
$$

This homogeneous system of equations admits solutions of the type (cf. eq. (6.36) and (6.67)),

$$
\mathbf{x}_1 = c(\tau, \rho)\mathbf{u}\, e^{ik_c R} + \mathrm{cc}
\qquad (6.104)
$$

where we have taken into account that $e^{ik_c R}$ is the (critical) eigenfunction of $\partial^2/\partial R^2$ in an infinite system (with periodic boundary conditions), $-k_c^2$ being the critical eigenvalue.

To $O(\varepsilon^2)$ one finds (cf. (6.89))

$$
\left[\mathscr{L}_0(\lambda_c) + \mathscr{L}_1(\lambda_c)\frac{\partial^2}{\partial R^2} \right]\cdot\mathbf{x}_2 = -\frac{1}{2}\mathbf{h}_{\mathbf{xx}}\cdot\mathbf{x}_1\mathbf{x}_1 - 2\mathscr{L}_1(\lambda_c)\frac{\partial^2 x_1}{\partial R\partial\rho}
$$

$$
(6.105)
$$

Unlike in the previous section, eq. (6.89), the solvability condition is not trivially satisfied since the last term in the right hand side still contains the critical eigenfunction e^{ik_cR}. One has to impose, therefore, that this term be orthogonal to the null eigenspace of the adjoint linear operator acting on x_2. We do not carry out explicitly the, by now familiar, manipulations and merely quote the final result, which is nothing but the requirement that $(\partial\omega/\partial k)_c = 0$. This condition is identically satisfied owing to the very nature of the criticality. We may therefore proceed to write the formal solution of (6.105) as

$$x_2 = \mathbf{p}_0|c|^2 + \mathbf{p}_1\frac{\partial c}{\partial\rho}e^{ik_cR} + \mathbf{p}_2c^2\,e^{2ik_cR} + \mathbf{p}_1^*\frac{\partial c^*}{\partial\rho}e^{-ik_cR} + \mathbf{p}_2^*c^{*2}\,e^{-2ik_cR}$$

$$(6.106)$$

where $\mathbf{p}_0$, $\mathbf{p}_1$, $\mathbf{p}_2$ are determined from (6.105).

To $O(\varepsilon^3)$ one finds the equation (cf. (6.91))

$$\left[\mathscr{L}_0(\lambda_c) + \mathscr{L}_1(\lambda_c)\frac{\partial^2}{\partial R^2}\right]\cdot x_3 = -\gamma_2\left[\mathscr{L}_{0\lambda}(\lambda_c) + \mathscr{L}_{1\lambda}(\lambda_c)\frac{\partial^2}{\partial R^2}\right]\cdot x_1$$

$$-\mathbf{h}_{xx}(\lambda_c)\cdot x_1 x_2 - \frac{1}{6}\mathbf{h}_{xxx}\cdot x_1 x_1 x_1 + \frac{\partial x_1}{\partial\tau} - \mathscr{L}_1(\lambda_c)\cdot\frac{\partial^2 x_1}{\partial\rho^2}$$

$$-2\mathscr{L}_1(\lambda_c)\cdot\frac{\partial^2 x_2}{\partial R\partial\rho} = \mathbf{q}_3 \qquad\qquad (6.107)$$

The solvability condition of this inhomogeneous system of equations requires

$$\int_0^{2\pi/k_c} dR\,e^{-ik_cR}\mathbf{u}^{+*}\cdot\mathbf{q}_3(c,\tau,R) = 0 \qquad\qquad (6.108)$$

where we used the same extended definition of scalar product as in Section 6.7 and $\mathbf{u}^+$ is determined by a procedure analogous to that leading to (6.72) and (6.73). Owing to the integration in R only those terms of $\mathbf{q}_3$ which depend on R through e^{ik_cR} will survive in (6.108). After some formal manipulations similar to those performed repeatedly in this chapter as well as in Chapter 5 one finally ends up with an equation for the normalized amplitude $z = \varepsilon c$ of the form

$$\partial z/\partial t = (\lambda - \lambda_c)P_1 z + D\,\partial^2 z/\partial r^2 - P_3|z|^2 z \qquad\qquad (6.109)$$

This equation differs from eqs. (6.94) and (6.95) by the important fact that the coefficients P_1, D and P_3 are *real-valued*, a consequence of the absence of an imaginary part in the critical eigenvalue and the corresponding eigenvector. (Actually, assuming supercritical bifurcation all three coefficients can be normalized to unity through an appropriate rescaling of z, r and t.) This makes (6.109) structurally identical to the Landau–Ginzburg equation which appears in the study of equilibrium critical phenomena. It is to be stressed, however, that the physics behind the two equations is very different. In particular, the 'potential' from which (6.109) derives in the sense of

$$\partial z/\partial t = -\delta U/\delta z^* \qquad (6.110a)$$

is given by

$$\mathbf{U} = \int dr \left[-(\lambda - \lambda_c)|z|^2 + \left|\frac{\partial z}{\partial r}\right|^2 + \frac{1}{2}|z|^4 \right] \qquad (6.110b)$$

It has a kinetic origin connected to the dynamics, in contrast to the classical Landau–Ginzburg equation which derives from the equilibrium free energy functional.

Let us now turn to the extension of (6.109) in two dimensions. We are interested in situations (cf. Sections 6.3 and 6.4) in which the basic structure arising near bifurcation is one-dimensional (roll type of pattern), along say the x-coordinate. The explicit form of the linearized dispersion relations (eq. (6.33a) or (6.42b)) then reads (Manneville, 1991)

$$\omega \approx a(\lambda - \lambda_c) - b[(k_x^2 + k_y^2)^{1/2} - k_c]^2$$

$$\approx a(\lambda - \lambda_c) - b\{[(k_c + \Delta k_x)^2 + \Delta k_y^2]^{1/2} - k_c\}^2$$

$$\approx a(\lambda - \lambda_c) - b[(k_c^2 + 2k_c\Delta k_x + \Delta k_x^2 + \Delta k_y^2)^{1/2} - k_c]^2$$

$$\approx a(\lambda - \lambda_c) - \frac{b}{4k_c^2}(2k_c\Delta k_x + \Delta k_x^2 + \Delta k_y^2)^2 \qquad (6.111)$$

Comparing orders of magnitude on both sides and bearing (6.51) in mind one realizes that

$$\Delta k_x \sim \Delta k_y^2$$

or

$$\partial^2 z/\partial x^2 \sim \partial^4 z/\partial y^4$$

This leads us to extend the scaling of space coordinates in eq. (6.60b), in the following way,

$$\frac{\partial}{\partial x} = \frac{\partial}{\partial X} + \varepsilon \frac{\partial}{\partial x_1} + \cdots$$

$$\frac{\partial}{\partial y} = \varepsilon^{1/2} \frac{\partial}{\partial y_1} + \cdots \qquad (6.112)$$

Replacing the last of eqs. (6.102) by (6.112) and again following the procedure described above one arrives at

$$\frac{\partial z}{\partial t} = (\lambda - \lambda_c)P_1 z + D\left(\frac{\partial}{\partial x} - \frac{i}{2k_c}\frac{\partial^2}{\partial y^2}\right)^2 z - P_3|z|^2 z \quad (6.113)$$

This equation, referred to as the *Newell–Whitehead–Segel equation*, can be regarded as the normal form of a symmetry-breaking bifurcation leading to roll or stripe patterns and allowing for modulation in both x and y directions. Fig. 6.7 illustrates on the Brusselator model, the type of pure and modulated patterns that may arise under these conditions. Note that when basic structures other than rolls are considered this equation needs to be corrected further, since one has then to allow for quadratic terms (Walgraef *et al.*, 1982).

B Secondary bifurcations. The Eckhaus instability

To get the flavor of the type of behavior predicted by the equations derived in the previous subsection we consider eq. (6.109) with the scaling $P_1 = D = P_3$,

$$\frac{\partial z}{\partial t} = (\lambda - \lambda_c)z + \frac{\partial^2 z}{\partial r^2} - |z|^2 z \quad (6.114)$$

One can easily check that this equation admits a family of time-independent solutions of the form

$$z_s = (\lambda - \lambda_c - \Delta k^2)^{1/2}\, e^{i(\Delta k r + \phi)} \quad (6.115)$$

where Δk is a small wave number describing the modulation of the basic structure at the critical wavelength k_c, and ϕ is an arbitrary phase. To study the stability of these solutions we set

Fig. 6.7 Numerical solutions of Turing patterns in the Brusselator model (eq. (3.28) supplemented with diffusion) in a two-dimensional system with periodic boundary conditions: (a) regular roll pattern obtained for the dimensionless variables $A = 4.5$, $B = 10$, $D_1 = 7$, $D_2 = 56$, $L_x = L_y = 64$; (b) irregular pattern displaying defects obtained for the same chemical and diffusion parameters as in (a) but with $L_x = L_y = 256$ and random initial conditions (De Wit, 1993).

(a)

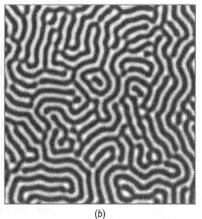

(b)

$$z = (\lambda - \lambda_{\mathrm{c}} - \Delta k^2)^{1/2} [1 + \delta z(r, t)] \, e^{i(\Delta k r + \phi)} \qquad (6.116)$$

The linearized equation for δz reads

$$\frac{\partial \delta z}{\partial t} = - (\lambda - \lambda_{\mathrm{c}} - \Delta k^2)(\delta z + \delta z^*) + 2i\Delta k \frac{\partial \delta z}{\partial r} + \frac{\partial^2 \delta z}{\partial r^2} \quad (6.117)$$

As for eq. (6.98) we decompose δz in real and imaginary parts, $\delta z = u + iv$, thus transforming (6.117) into

$$\left.\begin{aligned}
\frac{\partial u}{\partial t} &= -2(\lambda - \lambda_{\mathrm{c}} - \Delta k^2)u - 2\Delta k \frac{\partial v}{\partial r} + \frac{\partial^2 u}{\partial r^2} \\[2mm]
\frac{\partial v}{\partial t} &= 2\Delta k \frac{\partial u}{\partial r} + \frac{\partial^2 v}{\partial r^2}
\end{aligned}\right\} \quad (6.118)$$

We seek solutions of these equations in the form

$$\begin{pmatrix} u \\ v \end{pmatrix} = \begin{pmatrix} u_0 \\ v_0 \end{pmatrix} e^{i\kappa r} \, e^{\omega t}$$

Eqs. (6.118) are then transformed into a set of homogeneous algebraic equations for u_0, v_0 whose characteristic equation turns out to be

$$\omega^2 + 2[(\lambda - \lambda_{\mathrm{c}} - \Delta k^2) + \kappa^2]\omega + [2(\lambda - \lambda_{\mathrm{c}} - \Delta k^2) + \kappa^2]\kappa^2 - 4\kappa^2 \Delta k^2 = 0$$

Assuming κ is small we may neglect quartic terms in κ. The constant term then becomes equal to

$$2\kappa^2(\lambda - \lambda_{\mathrm{c}} - 3\Delta k^2) + O(\kappa^4)$$

entailing that an instability associated with a change of sign of ω will develop for wave numbers Δk such that

$$|\Delta k| > \left(\frac{\lambda - \lambda_{\mathrm{c}}}{3}\right)^{1/2} \qquad (6.119)$$

In other words, periodic patterns with modulation characterized by a wavelength sufficiently far from the basic structure ($\Delta k = 0$) are unstable with respect to long wavelength modes. This is known as the Eckhaus instability (Eckhaus, 1965). Evidence for the existence of this instability has been reported in fluid dynamics (Lowe and Gollub, 1985).

If perturbations transverse to the basic structure are allowed then one has to appeal to eq. (6.113). The analysis shows (Coullet and Gil, 1988; Manneville, 1991) that for negative Δk an instability with respect to large-scale wavy perturbations transverse to the pattern, referred to as the *zig-zag instability*, takes place. This transition introduces a deformation of the basic pattern and renders it two-dimensional. It has been observed experimentally in connection with the Bénard problem (Busse and Whitehead, 1971; Busse, 1978).

It is worth noting that the Eckhaus and zig-zag instabilities provide a partial solution of the selection problem raised in Section 6.5, since they imply the collapse of certain types of structure that would appear to be allowed on the basis of linear stability analysis, see Fig. 6.8.

C Phase dynamics

A very interesting array of behaviors allowed by the normal form equations derived in Section 6.8 and in the previous two subsections is revealed by recasting these equations into a form exhibiting an amplitude and a phase variable. To be specific, consider the complex Landau–Ginzburg equation (6.95) in a one-dimensional system and set

$$z = A(r, t)\, e^{i\phi(r, t)} \qquad (6.120)$$

Substituting into eq. (6.95) and separating real and imaginary parts one obtains

$$\frac{\partial A}{\partial t} = \left[(\lambda - \lambda_c) - \left(\frac{\partial \phi}{\partial r}\right)^2 - \alpha \frac{\partial^2 \phi}{\partial r^2} \right] A - A^3 + \frac{\partial^2 A}{\partial r^2} - 2\alpha \frac{\partial A}{\partial r} \frac{\partial \phi}{\partial r} \qquad (6.121a)$$

$$\frac{\partial \phi}{\partial t} = -\beta A^2 + \frac{\partial^2 \phi}{\partial r^2} - \alpha \left(\frac{\partial \phi}{\partial r}\right)^2 + \frac{\alpha}{A} \frac{\partial^2 A}{\partial r^2} + \frac{2}{A} \frac{\partial A}{\partial r} \frac{\partial \phi}{\partial r} \qquad (6.121b)$$

In the absence of space dependences these equations reduce to (5.55). They show that, while the amplitude A reaches a plateau value through an intrinsic relaxation mechanism with a well-defined characteristic time

Fig. 6.8 Mode selection arising from the Eckhaus and zig-zag instabilities: (a) marginal stability line; (b) Eckhaus instability line; (c) zig-zag instability line. The shaded area denotes the region of allowed modes.

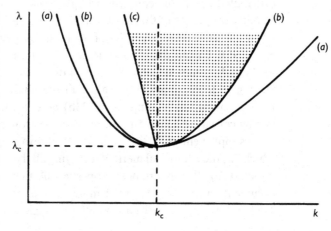

associated with the presence of a linear term $(\lambda - \lambda_c)A$, the phase ϕ follows the variations of A passively and, to the extent that A is small close to the bifurcation, it varies slowly in time. Coming back now to the full space-dependent equations this suggests that A can be eliminated and an autonomous dynamics for ϕ can be derived (except for a very narrow region around the instability threshold in which A will be subjected to critical slowing down), provided that the spatial derivatives of ϕ and A in the right hand side of eq. (6.121b) remain small.

The above envisaged reduction can be carried out most easily under the more stringent assumption that A is space-independent. Neglecting the time derivative of A in the first equation (6.121a) (a legitimate procedure after an initial time layer, according to the previous argument) one obtains an expression for A in terms of the phase,

$$A^2 = (\lambda - \lambda_c) - \left(\frac{\partial \phi}{\partial r}\right)^2 - \alpha \frac{\partial^2 \phi}{\partial r^2} \qquad (6.122)$$

Substituting into eq. (6.121b) one obtains

$$\frac{\partial \phi}{\partial t} = -\beta(\lambda - \lambda_c) + (1 + \alpha\beta)\frac{\partial^2 \phi}{\partial r^2} + (\beta - \alpha)\left(\frac{\partial \phi}{\partial r}\right)^2$$

or, introducing the new phase variable

$$\psi = \phi + \beta(\lambda - \lambda_c)t \qquad (6.123a)$$

$$\frac{\partial \psi}{\partial t} = (1 + \alpha\beta)\frac{\partial^2 \psi}{\partial r^2} + (\beta - \alpha)\left(\frac{\partial \psi}{\partial r}\right)^2 \qquad (6.123b)$$

This *nonlinear diffusion equation* turns out to have a structure identical to the Burgers equation encountered in hydrodynamics (Burgers, 1948; Ortoleva and Ross, 1973; Kuramoto, 1984). It describes nicely the spatial complexity arising from the desynchronization of an array of spatially distributed oscillators coupled through diffusion, as a result of the weak stability properties of the phase variable. Now, as we saw in Section 6.8, when $1 + \alpha\beta < 0$ an instability of the basic limit cycle solution occurs in eq. (6.95). Under these conditions the 'diffusion coefficient' of ψ in (6.123b) becomes negative. An 'anti-diffusion' behavior then sets in, leading to increasingly large values of the phase gradient. Clearly, the basic assumptions leading to (6.123b) need to be revised. A phenomenological way to achieve a saturation of the growth of the phase gradient is to supplement (6.123b) by a fourth derivative term, $-K(\partial^4\psi/\partial r^4)$ which has the additional merit of satisfying all the symmetry requirements imposed by the invariance properties of the initial equations. The argument can be justified by a more systematic procedure (Kuramoto 1984, 1990). Eq. (6.123b) augmented by such a fourth order derivative term is known as the *Kuramoto–Shivashinski equation*. It gives rise to a very rich behavior including the possibility of spatio-temporal chaos.

Notice that diffusion-like phase equations can only arise in dissipative systems under nonequilibrium constraints: in nondissipative systems the phase variable necessarily has a propagative character.

Phase equations associated with other types of instabilities can also be derived (Walgraef, 1988; Brand, 1990). We do not develop the technicalities here, but refer the reader to the original literature. We close this chapter by noticing that the phase dynamics formalism provides a very interesting insight into the origin of the defects characterizing nonequilibrium structures, referred to repeatedly in Chapter 1 and illustrated again in connection with Fig. 6.6. The latter appear as singular solutions of the normal form equations in which the amplitude variable A vanishes at the 'core' and the phase gradient has a circulation of $\pm 2\pi$ around any path, however small, surrounding the core (Coullet and Gil, 1988).

Problems

6.1 Perform linear stability analysis of the Brusselator (eq. (2.44)) in a square and in a circular spatial domain using either B or the size of the domain as the control parameter. Identify possible high-codimension instabilities arising from symmetry (Erneux and Herschkowitz-Kaufman, 1975).

6.2 Determine the conditions under which the Brusselator in a small one-dimensional box admits a codimension 2 instability originating from the coalescence of two time-independent modes of wave numbers m and $m + 1$. Derive the solutions associated with the secondary bifurcation close to this critical situation (Mahar and Matkowsky, 1977).

6.3 Compute the coefficients of the normal form equations (6.95) and (6.109) for the Brusselator in one space dimension, respectively near its Hopf and its Turing bifurcation.

$\Bigg($Answer: For (6.95) before rescaling (compare also with Appendix A2)

$$\frac{\partial z}{\partial t} = \frac{B - B_c}{2} z - \left(\frac{A^2 + 2}{2A^2} + \frac{\mathrm{i}}{2}\frac{4A^4 - 7A^2 + 4}{3A^3}\right)|z|^2 z + \frac{1}{2}[(D_1 + D_2) + \mathrm{i}A(D_2 - D_1)]\frac{\partial^2 z}{\partial r^2}$$

For (6.109)

$$\frac{\partial z}{\partial t} = \frac{1 + A(D_1/D_2)^{1/2}}{1 - D_1/D_2}\frac{B - B_c}{B_c} z - \frac{-8A^3(D_1/D_2)^{3/2} + 5A^2(D_1/D_2) + 38A(D_1/D_2)^{1/2} - 8}{9A^3(D_1/D_2)^{1/2}(1 - D_1/D_2)}|z|^2 z$$

$$+ \frac{4D_1[1 + A(D_1/D_2)^{1/2}]}{B_c(1 - D_1/D_2)}\frac{\partial^2 z}{\partial r^2}\Bigg)$$

6.4 Pattern formation in the presence of a preexisting shallow gradient (Gierer and Meinhardt, 1972; Almirantis and Nicolis, 1987). Consider a two-component reaction–diffusion system

$$\partial X/\partial t = A + f(X, Y, \lambda) + D_1 \nabla^2 X$$

$$\partial Y/\partial t = g(X, Y, \lambda) + D_2 \nabla^2 Y$$

in which the source term A exhibits a shallow gradient in space, $A = A_0 + \varepsilon a(\mathbf{r})$, $\varepsilon \ll 1$. Formulate the stability and bifurcation analyses in the presence of such a gradient and derive the extended normal form equation (in a small box) close to the Turing bifurcation. Show that in one space dimension the presence of the gradient entails a shift of the bifurcation point or the destruction of the bifurcation, in the sense of Fig. 5.5(b).

6.5 Pattern formation in the presence of anisotropies. In many systems like, for instance, nematic liquid crystals subject to elliptical shear (Guazzelli and Guyon, 1982) or condensed matter under irradiation (Martin, 1983) there exists an intrinsic anisotropy breaking the symmetry between two transverse orientation axes. Perform the stability analysis of the homogeneous state in a two-variable reaction–diffusion system in which anisotropy is accounted for by two different longitudinal and transverse diffusion coefficients $D_\parallel$ and D_T respectively. Apply the general formulae to the Brusselator model (Dewel, Borckmans and Walgraef, 1984).

6.6 The effect of imperfections in the Bénard instability (Ahlers, Hohenberg and Lücke, 1984; Lücke and Schank, 1985). Derive the extended version of the Boussinesq equations (Section 2.5) and of the corresponding normal form near the Bénard instability threshold for a roll pattern (Section 6.7) in the presence of a slight sustained periodic variation of temperature at the lower boundary of the layer. Show that the imperfection induces a displacement of the critical Rayleigh number in the direction of increased stability.

6.7 Bifurcations in nonideal reaction–diffusion systems (Li *et al.*, 1981; Li and Nicolis, 1981; Othmer, 1976). Derive the extended form of reaction–diffusion equations for the second Schlögl model (eq. (2.45b)) and for the Brusselator (eq. (2.44)) using the regular solution model in which the excess free energy is given by

$$G_e = \sum_{i \le j} w_{ij} n_i n_j / n$$

where $\{n_i\}$ are the mole numbers and w_{ij} account for the interactions. By further assuming that only the interactions between the intermediates (X, ...) and the initial products (A,B, ...) are nonideal, prove that the nonideality correction to the diffusion coefficients is of the form $D(X) = D_0(1 - 2wX)$ etc. Investigate the effect of these corrections near the cusp (for the Schlögl model) and the symmetry-breaking (for the

Brusselator) instabilities, with special emphasis on the interference between such instabilities and the phase transition (unmixing) predicted by the regular solution model.

6.8 The normal form of a dynamical system in the vicinity of the criticality (Re ω_c = Im ω_c = 0, k_{m_c} = 0) is given by an equation similar to (6.109) in which the order parameter z is real. By redefining space, time and z scales this equation can be written in one space dimension and in the presence of an imperfection (in the sense of Section 5.6) in the form (assuming $P_3 > 0$)

$$\frac{\partial z}{\partial t} = -z(z-a)(z-b) + \frac{\partial^2 z}{\partial r^2}$$

Check that in an unbounded system this equation admits for $b = 1$, $0 < a < 1$ *solitary wave* solutions of the form

$$z(r, t) = z(r - vt) = \frac{1}{1 + \exp[(1/\sqrt{2})(r - vt)]}$$

where the propagation velocity is $v = (1/\sqrt{2})(1 - 2a)$.

For a = 1/2 these solutions reduce to a stationary state (kink) joining states $z = 0$ and $z = 1$. Derive this solution directly from the normal form by setting $\partial z/\partial t = 0$ and treating the resulting equation by methods analogous to those of Chapters 3–5, in which time is now replaced by the spatial coordinate r (Campbell, Newell, Schrieffer and Segur, 1986; Malchow and Schimansky-Geier, 1985).

6.9 First order phase transitions in systems under constraint (Langer, 1980; Langer and Müller-Krumbhaar, 1983). In a great number of situations a liquid–solid phase transition takes place when the solidification front advances under the action of an external nonequilibrium constraint. A model equation describing dendritic crystallization under such conditions is

$$\frac{\partial R}{\partial t} = -v\frac{\partial R}{\partial x} - R\frac{\partial^2 R}{\partial x^2} - \frac{\partial^4 R}{\partial x^4}$$

where R is the radius of curvature at the tip, and v the velocity of the front. The minus sign of the second order derivative term accounts for phase instability (negative diffusion) whereas the fourth order derivative term describes stabilization through surface effects. Perform a linear stability analysis of this equation around the reference state $R(x, t) = r_0$ = const. and identify the characteristic lengths and time scales present in the problem. Does the equation admit bounded solutions in the fully nonlinear regime?

6.10 Precipitate pattern formation. An alternative interesting mechanism of interference between phase transitions and nonequilibrium constraints is provided by the competitive particle growth model (Feinn *et al.*, 1978).

Let c represent a monomer concentration, $R(x, t)$ the 'local' radius of particles formed by monomer aggregation. A minimal model describing this process of precipitation is

$$\frac{\partial R}{\partial t} = k[c - c^{eq}(R)]$$

$$\frac{\partial c}{\partial t} = D\nabla^2 c - 4\pi npR^2 \frac{\partial R}{\partial t} + W$$

where $c^{eq}(R)$ is an equilibrium value of c for a radius R particle, n the particle number density, p the molar density and W the rate of monomer production (if any). Perform a linear stability analysis of the uniform state $R = r_0$ (arbitrary), $c = c^{eq}(r_0)$. Identify the type of instability that can be realized, particularly in connection with the existence of a characteristic length of intrinsic origin.

CHAPTER SEVEN

Chaotic dynamics

7.1 The Poincaré map

As we have seen throughout this monograph, in a nonlinear dynamical system the first bifurcation from a fixed point leads to fixed point or to limit cycle behavior. Chaotic behavior, which according to the experimental data surveyed in Chapter 1 is abundant in nature, can therefore arise smoothly from simple fixed point behavior only through a sequence of bifurcations involving high order (tertiary etc.) transitions. As a rule, at some stage of this sequence of transitions a periodic solution loses its stability, a fact that is also reflected in the experimental data where chaotic behavior seems to be much more intimately intertwined with periodic rather than steady-state behavior.

The above comments suggest that to gain an insight into the onset of chaos it is necessary to analyze the loss of stability and the subsequent bifurcation behavior of periodic solutions. Unfortunately, this task is unattainable. First, the analytic form of these solutions in the interesting parameter region is not known except in a number of exceptional situations. Second, even if the analytic form were known one would be led to study dynamical systems of the form of eq. (3.26) in which both the linearized operator $\mathscr{L}$ and the nonlinear part $\mathbf{h}$ contain an explicit periodic dependence in time. This is, in principle, possible by Floquet theory (Cesari, 1963; Hale and Koçak, 1991), but in practice it is much more difficult to carry out on a quantitative basis than is the study of the behavior around fixed points. The study of higher order transitions is, of course, even more complicated.

An elegant way out of this difficulty has been invented by Poincaré. We illustrate the idea in the case of a three-dimensional phase space (Fig. 7.1). Let γ be the phase space trajectory of a dynamical system and S a surface cutting this trajectory transversally. The trace of γ on S is a sequence of points $P_0, P_1, P_2, \ldots$ at which the trajectory intersects the surface with a slope of prescribed sign. The successive positions $\mathbf{x}_0, \mathbf{x}_1, \mathbf{x}_2, \ldots$ of these points follow, in principle, from the dynamics. Indeed, if we write the formal solution of (3.26) as

$$\mathbf{x}_t = \phi_t(\mathbf{x}_0) \tag{7.1}$$

then, clearly, $\mathbf{x}_1 = \phi_{T(\mathbf{x}_0)}(\mathbf{x}_0)$ where $T(\mathbf{x}_0)$ is the time necessary for the trajectory to return to S with the same sign of slope starting from $\mathbf{x}_0$, and similarly $\mathbf{x}_2 = \phi_{T(\mathbf{x}_1)}(\mathbf{x}_1)$ etc. Quantitatively the precise form of ϕ and the values of $T(\mathbf{x}_i)$ are impossible to determine. But if we label the successive points not by the time at which they are visited but, rather, by their order, then we realize that the original flow (7.1) induces on S a dynamics of the form

$$\mathbf{x}_{n+1} = \mathbf{g}(\mathbf{x}_n) \tag{7.2}$$

where $\mathbf{g}(\mathbf{x}) = \phi_{T(\mathbf{x})}(\mathbf{x})$. This dynamics is no longer continuous in time: it is a *recurrence*, since the time intervals between successive intersections are finite. This is referred to as the *Poincaré map*, S being the *Poincaré surface of section*. By introducing an appropriate coordinate system in S, we can write (7.2) in the more explicit form

$$\left.\begin{aligned} x_{n+1} &= f(x_n, y_n) \\ y_{n+1} &= g(x_n, y_n) \end{aligned}\right\} \tag{7.3}$$

We notice that, starting with a three-dimensional dynamical system we have ended up with a two-dimensional recurrence. This reduction of dimensionality is one of the important advantages of the Poincaré map. The second advantage is that recurrences like eqs. (7.2) and (7.3) lend

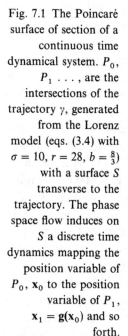

Fig. 7.1 The Poincaré surface of section of a continuous time dynamical system. P_0, $P_1 \ldots$, are the intersections of the trajectory γ, generated from the Lorenz model (eqs. (3.4) with $\sigma = 10, r = 28, b = \frac{8}{3}$) with a surface S transverse to the trajectory. The phase space flow induces on S a discrete time dynamics mapping the position variable of P_0, $\mathbf{x}_0$ to the position variable of P_1, $\mathbf{x}_1 = \mathbf{g}(\mathbf{x}_0)$ and so forth.

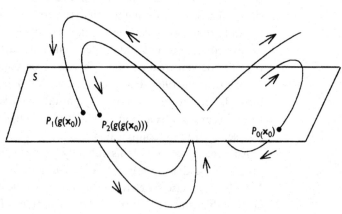

themselves to numerical simulation much better and much more accurately than continuous time dynamical systems. A third advantage is that by determining the object that will attract the P_ns for long times, we will be able to infer a number of key properties of the attractor of the original dynamical system, since in actual fact we will dispose of a section (rather than a projection) of this attractor. In particular, the stability properties of these two objects will be completely identical.

Let us give some illustrations of this last point. By construction, a limit cycle attractor will intersect the Poincaré surface of section with a given slope at a single point P which will remain invariant under the dynamics, i.e. for all successive intersections. At the level of eqs. (7.2) and (7.3) this will be reflected by the fact that the coordinates $(\bar{x}, \bar{y})$ of P will be such that $\bar{x}_{n+1} = \bar{x}_n, \bar{y}_{n+1} = \bar{y}_n$ or

$$\left. \begin{aligned} \bar{x} &= f(\bar{x}, \bar{y}) \\ \bar{y} &= g(\bar{x}, \bar{y}) \end{aligned} \right\} \quad (7.4)$$

In other words, a periodic solution corresponds to a *fixed point* of the recurrence generated on S. An interesting generalization is to search for attractors which produce on S a finite sequence of points visited consecutively such that $x_{n+1}, \ldots, x_{n+s}$ remain different from x_n (similarly for y) until an iteration k is reached for which

$$\left. \begin{aligned} x_{n+k} &= x_n \\ y_{n+k} &= y_n \end{aligned} \right\} \quad (7.5)$$

We call this a cycle of order k. Fig. 7.2 describes how a cycle of order two

Fig. 7.2 A cycle of order two and its signature on the Poincaré surface of section.

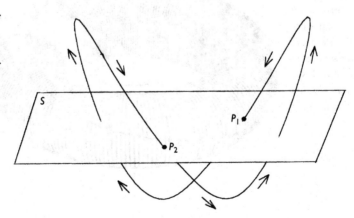

looks in the original phase space. We notice (see also Section 3.3C) that the additional twisting of the trajectory before closing to itself would be impossible in a two-dimensional phase space as it would imply self-intersection.

Another topologically interesting possibility, depicted in Fig. 7.3, arises when the intersection points of the trajectories converge to a closed curve C in the Poincaré surface of section. Following the discussion in Section 3.3 the corresponding attractor of the flow is a *torus*.

In the above perspective the signature of chaotic dynamics on the Poincaré surface of section should be a set that is equivalent to neither a countable set of points nor a smooth curve. We have encountered such objects, the *fractals*, in our analysis of invariant manifolds of Section 3.3. Their existence entails that the underlying attractor of the continuous time dynamical system should contain an uncountable number of sheets whose transverse intersection by a line produces a Cantor-like set. This, in turn, requires that the attractor undergoes successive *foldings* as time goes on (cf. Fig. 3.7).

Fig. 7.4(a) depicts the Poincaré surface of section of the Rössler attractor (eqs. (4.41) and Fig. 4.10) on a surface of section corresponding to the plane $y = 0, x < 0, z < 1$ (Gaspard and Nicolis, 1983). We obtain a cloud of points. By delimiting part of the surface of section by a small rectangle we observe that at later intersections this rectangle is rotated and deformed into a stick-like structure which, in turn, is folded into a horseshoe-like structure. Subsequent foldings will produce the Cantor-like structure anticipated above.

A second view of chaos in the Rössler model is given by Fig. 7.4(b). Here

Fig. 7.3 A quasi-periodic attractor in the form of a two-dimensional torus and its signature on the Poincaré surface of section in the form of a closed curve C (after Thompson, 1982).

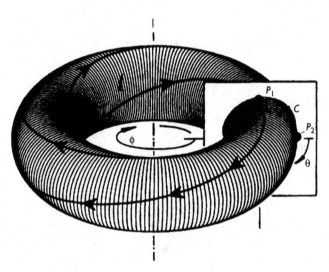

Fig. 7.4 (*a*) Horseshoe
map induced by
Rössler's model (eqs.
(4.41)) on the Poincaré
surface of section
($y = 0$, $x < 0$, $z < 1$).
The trajectories within
the rectangle rotate
around the
one-dimensional stable
manifold of the fixed
point and intersect the
Poincaré surface after
having followed the
folding of the unstable
manifold. (*b*) The
one-dimensional map
obtained by plotting
the value of $-x$ at the
$(n + 1)$th intersection
of the trajectory with
the Poincaré surface,
versus its value at the
nth intersection.

(a)

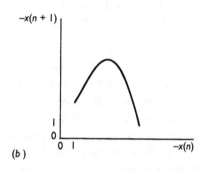

(b)

we plot the value of x at the $(n + 1)$th intersection point between the
above-defined Poincaré surface and the flow, as a function of its value at
the nth intersection. The numerical construction shows that we obtain a
smooth bell-shaped curve. The positions of the successive intersections on
this curve are not given by consecutive points but by points that appear to
be distributed randomly.

Fig. 7.5 depicts a second example of a *return map*, obtained this time
from a seven-variable model of the Belousov–Zhabotinski reaction
(Richetti and Arnéodo, 1985; see also Section 1.4). The temporal
variation of Br$^-$ion concentration and a two-dimensional projection of
the attractor of the model equations are represented, respectively, in Figs.
7.5(*a*) and 7.5(*b*). The return map itself, Fig. 7.5(*c*), is obtained from a
Poincaré surface of section whose trace is indicated by the dashed line on
Fig. 7.5(*b*). These results follow very closely the time dependences,
attractor shapes and return maps obtained directly from the experimental
data (Turner, Roux, McCormick and Swinney, 1981).

In both Fig. 7.4(*b*) and Fig. 7.5(*c*), the important thing is that in the
representation afforded by the return map one is left with a one-
dimensional recurrence, an additional advantage with respect to the
high-dimensional recurrence governing the full Poincaré surface of
section (Fig. 7.4(a)). Such dimensionality reductions arise frequently in
systems possessing widely separated time scales: as time goes on fast
processes associated with contraction wipe out any extension in certain
directions, leaving only one relevant variable associated with a slow
motion along the most unstable direction.

7.2 One-dimensional recurrences: general aspects

The detailed construction of the Poincaré recurrence on the surface of section, starting from a given dynamical system, is an extremely arduous task that cannot be carried out quantitatively unless the solutions of the equations of evolution (3.26) are known explicitly. As stressed repeatedly, in most dynamical systems of interest this is not the case. Nevertheless, the Poincaré map is the origin of a most fruitful approach to chaos which may be summarized as follows:

One starts with a particular recurrence, or a family thereof, arguing that there is bound to be a family of phase space flows reducible to this recurrence through a judicious choice of the surface of section.

Qualitative, and whenever possible, quantitative analysis is performed on the recurrence. Certain features concerning possible routes to chaos as well as fully developed chaos that would be impossible to unravel from the original flow are thus brought out.

By the previous argument on the existence of an underlying family of flows amenable to the Poincaré recurrence, it is conjectured that the conclusions obtained from the above analysis should be generic for flows as well, provided that the form of the evolution laws on the Poincaré surface is not pathological. The validity of this conjecture is checked by experiment or by numerical simulation on model equations.

Fig. 7.5 (*a*) Temporal variation of the Br^- ion concentration deduced from the seven-variable model of the Belousov-Zhabotinski reaction by Richetti and coauthors; (*b*) two-dimensional projection of the phase portrait generated by the model equations; (*c*) the return map deduced from a Poincaré surface of section cutting the attractor along a hyperplane whose trace is indicated by the dashed line in (*b*).

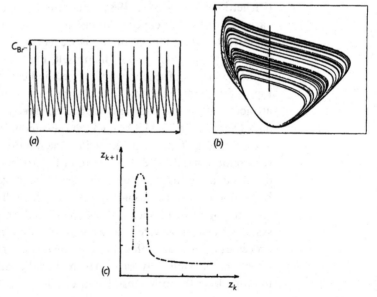

Experience acquired during the last decade has fully vindicated the interest in this approach to which, consequently, much of this chapter will be devoted. In this section we derive some general properties of recurrences (Collet and Eckmann, 1980; Hao, 1989). In view of the arguments developed at the end of Section 7.1 we focus on the one-dimensional case

$$x_{n+1} = f(x_n, \mu) \tag{7.6}$$

where μ stands for a control parameter.

Let $\bar{x}$ be a fixed point solution of (7.6),

$$\bar{x} = f(\bar{x}, \mu) \tag{7.7}$$

In the spirit of Section 3.5, we introduce a perturbation ξ_n from $\bar{x}$ through

$$x_n = \bar{x} + \xi_n \tag{7.8}$$

substitute in (7.6), expand f in Taylor series around $\bar{x}$ and keep only the linear contribution in ξ. One obtains in this way, using property (7.7),

$$\xi_{n+1} = f'(\bar{x}, \mu)\xi_n \tag{7.9a}$$

where the prime denotes the first derivative. This equation admits solutions of the form

$$\xi_n = c\rho^n \tag{7.9b}$$

Substitution into (7.9a) yields

$$\rho = f'(\bar{x}, \mu) \tag{7.10a}$$

On the other hand, from (7.9b) it is seen that ξ_n will decay in time only if $|\rho| < 1$. We thus arrive at the condition of asymptotic stability of the fixed point

$$|f'(\bar{x}, \mu)| < 1 \tag{7.10b}$$

The above analysis can be extended straightforwardly to cycles of order k. Let $\{x_1, \ldots, x_k\}$ be the set of points visited consecutively by the cycle. Since $x_2 = f(x_1, \mu), \ldots, x_k = f(x_{k-1}, \mu)$, it follows that each of the points $\{x_1, \ldots, x_k\}$ satisfies the property

$$x_i = \underbrace{f(f \ldots f(x_i))}_{k \text{ times}} \equiv f^{(k)}(x_i) \tag{7.11}$$

i.e. it is a fixed point of the kth iterate of the map. Using the chain rule for the differentiation of implicit functions the condition of asymptotic stability (eq. (7.10b)) then takes the form

$$|f'(x_1, \mu) \ldots f'(x_k, \mu)| < 1 \tag{7.12}$$

In studies involving one-dimensional recurrences μ and f are usually chosen and normalized in such a way that if the initial data x_0 are taken from a finite interval $I \equiv (a, b)$, then the various iterates $x_1, \ldots, x_n$ belong to the same interval. Therefore, the function f maps the interval I *into* itself in the sense that the iterates may not fill the entire interval. We refer to such systems as *endomorphisms*. An endomorphism with a smooth inverse is referred to as a *diffeomorphism*. Typically, one-dimensional endomorphisms leading to chaos are not invertible owing to the elimination of the motion along the contracting directions. In contrast to diffeomorphisms for such systems the past cannot be reconstructed in a unique fashion.

7.3 Phenomenology of one-dimensional recurrences: illustrations

Before we proceed to the quantitative study of the mechanisms leading to chaotic behavior and to the characterization of fully developed chaos we illustrate the properties of one-dimensional maps on three typical examples: the logistic, the circle and the intermittent map. These also happen to be the dynamical systems on which much of the early fundamental work on chaos theory has concentrated (Schuster, 1988; Bergé *et al*, 1984; Baker and Gollub, 1990).

A The logistic map

The specific form of eq. (7.6) associated with this dynamical system is

$$x_{n+1} = 4\mu x_n(1 - x_n), \qquad \begin{cases} 0 \leq \mu \leq 1 \\ 0 \leq x \leq 1 \end{cases} \tag{7.13}$$

Its genericity stems from the fact that any function possessing a nondegenerate extremum behaves around this extremum as does the right hand side of eq. (7.13) around $x = 1/2$.

The fixed point equation (7.7) possesses two solutions:

$x_{10} = 0$, which exists for all values of the parameter μ
$x_{11} = 1 - 1/4\mu$, which exists as long as $\mu > 1/4$

The stability of these fixed points can be assessed from eq. (7.10b), which for eq. (7.13) reads

$$|4\mu - 8\mu\bar{x}| < 1 \tag{7.14}$$

For $\bar{x} = x_{10} = 0$ this leads to condition $\mu < 1/4$ in other words, the trivial solution is stable only as long as it is the unique fixed point of the system. For $\bar{x} = x_{11} = 1 - 1/4\mu$, inequality (7.14) leads to $1/4 < \mu < 3/4$, where the left part of the inequality guarantees that x_{11} is in the unit interval.

Fig. 7.6 depicts the evolution of an initial condition x_0 induced by eq. (7.13) for values of μ slightly below 3/4. We observe that after a short transient the iterates of x_0 spiral around the nontrivial fixed point x_{11}, to which they eventually tend as $n \to \infty$. Notice that x_{11} is the intersection of the graph of $f(x)$ and of the bissectrix of the unit box. Having drawn the bissectrix the iteration is visualized most conveniently first by going vertically from x_0 to its image on the graph, then by shifting this image on the bissectrix horizontally, then by continuing once again (vertically) toward the graph of $f(x)$ and so forth.

Beyond the stability limit $\mu = \frac{3}{4}$ of the fixed point x_{11} a new solution in the form of a cycle of order two takes over, as illustrated in Fig. 7.7 for $\mu = 0.775$. We refer to this phenomenon as *period doubling*. This solution, which we denote by (x_{2-}, x_{2+}) can be constructed analytically using eq. (7.5),

$$x_{2+} = 4\mu x_{2-}(1 - x_{2-})$$

$$x_{2-} = 4\mu x_{2+}(1 - x_{2+})$$

or, after some elementary algebra,

$$x_{2\pm} = \frac{1}{8\mu}[1 + 4\mu \pm (16\mu^2 - 8\mu - 3)^{1/2}] \qquad (7.15)$$

Fig. 7.6 Evolution toward the stable fixed point in the logistic map for $\mu = 0.7$.

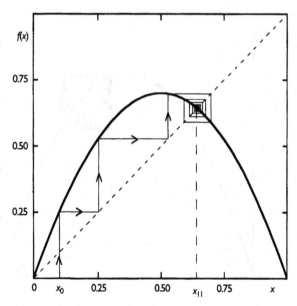

Referring to curve (*b*) of Fig. 5.1, we see that the two points of the cycle emerge from x_{11} through a mechanism of supercritical pitchfork bifurcation at the critical parameter value $\mu = 3/4$. Notice that at the level of the evolution induced by f the two values $x_{2\pm}$ are parts of a single solution, and are visited consecutively in the course of time. But at the level of the second iterate $f^{(2)}$ each of x_{2+}, x_{2-} corresponds to a distinct (fixed point) solution. It is, therefore, at this level that the analogy with pitchfork bifurcation is meaningful.

Let us jump, next, to the largest value of μ allowed by eq. (7.13), $\mu = 1$. One observes that for this value a typical initial condition in the interval will evolve to a complex, aperiodic behavior shown in Fig. 7.8(*a*),(*b*). As we shall see later, this behavior turns out to be one of the most clearcut and best-established examples of deterministic chaos.

The above discussion shows that the logistic map is capable of producing both regular, periodic behavior (Figs. 7.6 and 7.7) and irregular, aperiodic behavior (Fig. 7.8). The question therefore naturally arises, of whether these two types of regime are separated by a well-defined transition. We shall see in Section 7.5 that this is indeed the case. More specifically, the transition is manifested through an infinite sequence of successive period doublings at increasing values of the control parameter $\mu(\mu_1 = 3/4, \ \mu_2, \ldots, \ \mu_n, \ldots)$ culminating at a well-defined value $\lim_{n\to\infty}\mu_n = \mu_\infty < 1$.

The logistic map at $\mu = 1$ generates another, even simpler prototype of chaotic behavior through the change of variable

Fig. 7.7 Evolution
toward a stable cycle
of order two in the
logistic map for
$\mu = 0.775$.

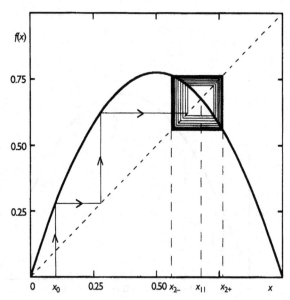

$$x_n = \sin^2 \frac{\pi y_n}{2} \tag{7.16a}$$

Substituting into (7.13) and using standard trigonometric identities one arrives at

$$\sin^2 \frac{\pi y_{n+1}}{2} = \sin^2 \pi y_n$$

implying that

$$\pi y_{n+1}/2 = \pi y_n \qquad \text{if} \quad 0 \le \pi y_n \le \pi/2$$

$$\pi y_{n+1}/2 = \pi - \pi y_n \qquad \text{if} \quad \pi/2 \le \pi y_n \le \pi$$

or finally

Fig. 7.8 Fully developed chaos in the logistic map for $\mu = 1$: (*a*) successive iterations starting from $x_0 = (\sqrt{5} - 1)/2$; (*b*) time series generated by these iterations.

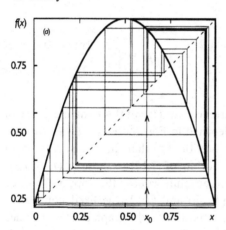

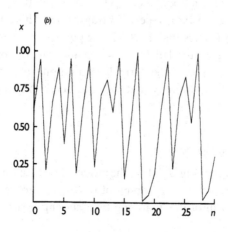

Fig. 7.9 The tent map
$T(x)$, eq. (7.16b).

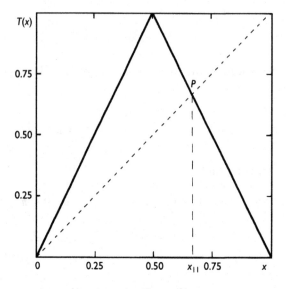

$$y_{n+1} = \begin{cases} 2y_n & 0 \leqslant y_n \leq \tfrac{1}{2} \\ 2 - 2y_n & \tfrac{1}{2} \leq y_n \leq 1 \end{cases} \qquad (7.16b)$$

The recurrence law $T(x)$ of this dynamical system, referred to as the *tent map*, is depicted in Fig. 7.9. It is *topologically conjugate* to (7.13), in the sense (cf. Section 3.6) that there exists a homeomorphism (eq. (7.16a)) taking the orbits of (7.13) to those of (7.16b).

Since the absolute value of the slope of $T(x)$ is everywhere larger than 1, by eqs. (7.10b) and (7.12) the fixed point and all periodic orbits are unstable. Inspection of the graph of $T^{(k)}(x)$ immediately shows that there are 2^k intersection points between this graph and the bissectrix. Of these, two are the fixed points of $T(x)$ itself ($x_{01} = 0$, $x_{11} = \tfrac{2}{3}$). The remaining $2^k - 2$ points belong to period k trajectories. Thus, there is one period-two trajectory (consisting of $2^2 - 2 = 2$ points), two period-three trajectories (consisting of a total of $2^3 - 2 = 6$ points) etc. By construction, the points belonging to these trajectories have a rational abscissa on the x-axis. One can prove additionally that they are dense in [0,1].

B The circle map

We have seen (Fig. 7.3) that the motion on a torus induces on the Poincaré surface of section a mapping of the circle into itself. Let us deduce the explicit form of this mapping by assuming first that the motion on the torus is uniform. In the angular coordinates θ and ϕ already introduced in Section 3.3C this gives rise to the following equations,

$$d\theta/dt = \omega_1$$

$$d\phi/dt = \omega_2$$

where ω_1, ω_2 are the (constant) angular velocities. The solution of these equations reads

$$\left. \begin{aligned} \theta(t) &= \theta_0 + \omega_1 t \\ \phi(t) &= \phi_0 + \omega_2 t \end{aligned} \right\} \quad (7.17)$$

Suppose that the Poincaré surface of section cuts the torus along the 'meridian' $\phi = 0 \pmod{2\pi}$. The intersection of the trajectories on the torus with this meridian will occur at times t_n such that $\phi(t_n) - \phi_0 = 2\pi n$ where n is an integer, or

$$t_n = 2\pi n/\omega_2 \qquad n = 1, 2, \ldots$$

Substituting into the first of eqs. (7.17) one finds

$$\theta_n = \theta_0 + 2\pi \frac{\omega_1}{\omega_2} n \qquad (\text{mod } 2\pi)$$

or, after normalizing and transforming to the usual form of a recurrence,

$$\theta_{n+1} = \alpha + \theta_n \qquad (\text{mod } 1) \qquad (7.18)$$

with

$$\alpha = \omega_1/\omega_2$$

This dynamical system is known as the *twist map* and is depicted in Fig. 7.10(a). If α is rational (a case which has already been referred to as *resonance*, see Section 2.1) the trajectory emanating from an initial point θ_0 will eventually return to θ_0 after a time $T = p/\omega_1 = q/\omega_2$ where p and q are integers such as $\omega_1/\omega_2 = p/q$. The corresponding trajectory on the original torus will be a closed curve winding p times along the θ direction and q times along the ϕ direction. But if α is irrational the trajectory will never return to its initial position: the motion will be quasi-periodic, and will be represented by a helix winding indefinitely on the torus. As most of the real numbers are irrational, this case will be typical for the twist map.

The twist map can be extended to account for nonuniform motion on the torus. The correction must, of course, respect the periodicity condition (eq. (7.18))

$$f(\theta + 1) = f(\theta) + 1$$

The most widely used model of $f(\theta)$ is the sine map

$$\theta_{n+1} = \theta_n + \alpha - (K/2\pi)\sin 2\pi\theta_n \qquad (\text{mod } 1) \qquad (7.19)$$

As the nonlinearity parameter K is tuned different types of mapping, depicted in Figs. 7.10(b) and (c), can arise. In case (b) $K < 1$, and the map is invertible since $f'(\theta) = 1 - K\cos 2\pi\theta$ can never vanish. In case (c) $K > 1$ and the map is noninvertible, resembling near its extrema the logistic map. This is at the origin of a variety of complex behaviors leading eventually to chaos.

Fig. 7.10 The circle map, eq. (7.19), for $\alpha = (\sqrt{5} - 1)/4$ and for increasing values of the nonlinearity parameter K: (a) $K = 0$ (twist map, eq. (7.18)); (b) $K = 0.9$; (c) $K = 2$; (d) state diagram in the parameter space (K, α). The motion is periodic inside the hatched regions (Arnol'd tongues) and quasi-periodic outside. At $K = 1$ the rational winding numbers dominate and for $K > 1$ the tongues overlap. (1)–(3) indicate various possible routes to chaos.

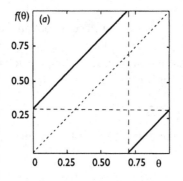

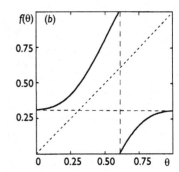

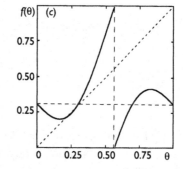

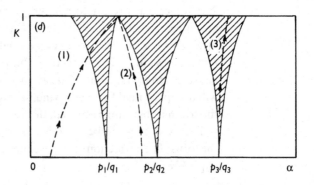

As in the logistic map one may inquire about the transition from the quasi-periodic behavior typical of (7.18) to the chaotic behavior generated from (7.19) when $K > 1$. Curiously, this transition may involve as an intermediate step regular regimes in the form of periodic oscillations. The way these regimes arise is depicted in Fig. 7.10(d). For $K \neq 0$ the regions of periodic behavior are no longer limited to the rational values of α, but correspond to whole regions (the so-called Arnol'd tongues) whose widths increase with K. Indeed, given a map like the one depicted in Fig. 7.10(b), there exists a sufficiently high iterate $f^{(k)}$ whose graph will intersect for a certain α the bissectrix at two points corresponding to one stable and one unstable cycle of order k. Since intersection between two curves is structurally stable with respect to slight changes of parameters, this situation is bound to subsist for a whole interval of values of α.

There exists an infinite number of such *phase-locked* intervals. Outside of these intervals the motion remains quasi-periodic. As α varies at fixed K, the map displays therefore both periodic and quasi-periodic behavior. But as K approaches 1, the rational intervals increase in size. At $K = 1$ the set of irrational intervals reduces to a fractal. Beyond this value the phase-locking regions overlap and several periodic oscillations can occur for given (K, α) depending on the initial conditions. Chaos is also observed for certain values of α. Three different paths are depicted in Fig. 7.10(d). They correspond, respectively, to the transitions quasi-periodicity→phase locking→chaos, (1); quasi-periodicity→chaos, (2); and simple periodicity→period doubling→chaos, (3).

C The intermittent map

A very interesting situation arises when one of the branches of the map is nearly tangent to the bissectrix. This case, which may actually be encountered in both the logistic and the circle map or their iterates in some parameter ranges, is depicted schematically in Fig. 7.11. After a short transient, an initial condition x_0 enters in the narrow region between the graph of the function and the bissectrix. At the beginning this process resembles convergence to a fixed point of marginal stability and displays, therefore, a very long time scale. But since the fixed point does not actually exist the iterates will eventually leave this region, evolve into the second branch of the graph of $f(x)$ near $x = 1$ and be reinjected shortly thereafter back to the region of near-tangency. To the observer this will appear as a series of long periods of quiescence interrupted at seemingly random times by short-lived bursts – a property which one usually refers to as *intermittency* and which is, indeed, one of the characteristic signatures of the phenomenon of turbulence.

From the above discussion it follows that a one-dimensional map possessing two fixed points which tend to merge and subsequently to disappear as a parameter is varied, can potentially show intermittency (Manneville and Pomeau, 1980). In the language of Section 5.5 this transition amounts, therefore, to a limit point bifurcation.

7.4 Tools of chaos theory

In much of the analysis carried out in Chapters 4–6 we have been able to formulate the onset of complex behavior as a *local* problem amenable to a perturbative approach. This has led us to normal form equations

Fig. 7.11 Generation of intermittent behavior through limit point bifurcation in the map $f(x) = 0.25 + \varepsilon + x^2$ mod 1: (*a*) $\varepsilon - 0.05$: the system possesses one stable and one unstable fixed point; (*b*) $\varepsilon = 0.02$: the fixed points have been destroyed and the system undergoes chaotic behavior of the intermittent type.

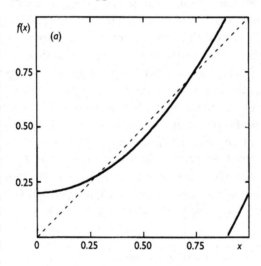

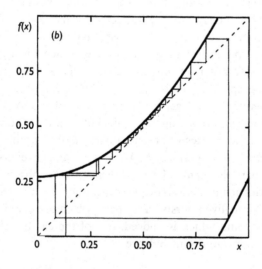

describing the spatio-temporal behavior of the amplitude of the bifurcating solutions in the vicinity of the transition point, which captured the essential part of the physics of the underlying problem.

As the discussion of the preceding sections has made clear, in contrast to the fixed points, limit cycles and spatial patterns, the onset and the principal properties of chaos constitute a global problem that is not amenable to perturbation theory. New approaches are needed. The present section is devoted to a brief survey of these methods, which will be applied subsequently in a number of case studies.

A first type of method stems from the idea, already implicit in the discussion of Section 3.4, that in the presence of complex dynamics the monitoring of a phase space trajectory in a pointwise fashion loses much of its interest. One attractive alternative to this limitation is *coarse-graining*: we partition the phase space into a finite number of cells $\{C_i\}$, $i = 1, \ldots, N$ (Fig. 7.12) and monitor the successive cell-to-cell transitions of the trajectory. One may look at the 'states' $C_1, \ldots, C_N$ as the symbols of an N-letter alphabet. In this view, then, the initial dynamics induces on the partition a *symbolic dynamics* describing how the letters of the alphabet unfold in time (Collet and Eckmann, 1980; Hao, 1989; Devanay, 1989). The investigation of this dynamics provides one with a powerful tool for classifying trajectories of various types and for unraveling aspects of the system's complexity that would remain blurred in a traditional description limited to the trajectories. Explicit examples will be given in Section 7.8. An additional motivation for developing the idea of symbolic

Fig. 7.12 Illustration of the idea of coarse-graining and symbolic dynamics.

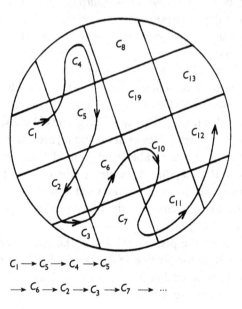

$$C_1 \longrightarrow C_5 \longrightarrow C_4 \longrightarrow C_5$$
$$\longrightarrow C_6 \longrightarrow C_2 \longrightarrow C_3 \longrightarrow C_7 \longrightarrow \cdots$$

dynamics is that in many natural phenomena strings consisting of sequences of letters play a central role. For instance, the DNA and RNA molecules are linear strings written on an alphabet consisting of four letters 'A', 'C', 'G', and 'T' (or 'U'), according to whether the nucleic acid subunit (nucleotide) contains the bases adenine, cytosine, guanine and thymine (or uracil). Furthermore most of the messages transporting information or having a cognitive value such as books, music, computer programs, or electrical activity of the brain are amenable in one way or the other to strings of letters.

Closely related to the ideas of coarse-graining and symbolic dynamics is the *statistical approach* to complex dynamical systems and especially to chaos (Lasota and Mackey, 1985; Eckmann and Ruelle, 1985). The objective here is to add to the topological view afforded by symbolic dynamics a *metric* element, such as the probability density $p_n(x)$ of finding the system in phase space point x at time n, introduced in Section 3.4, or (Nicolis and Nicolis, 1988) the probability $p_n(C_i)$ of finding the system in one of the cells of the partition of Fig. 7.12. In a similar vein one may also introduce joint probabilities and correlation functions. The interest and power of this description will be illustrated in Sections 7.6 and 7.8. Notice that, applied to conservative systems, the statistical description of chaos becomes intimately related to the foundations of statistical mechanics. In this chapter, however, we shall limit the analysis to dissipative chaos.

A third type of method for tackling chaos stems from the observation obtained from numerical studies that, in many instances, the dynamics exhibits a remarkable *self-similarity*. This property, which is also at the basis of the description of chaos by fractal attractors (cf. Section 3.3D), is nicely illustrated on the logistic map introduced in Section 7.3A (Devanay, 1989).

In Figs. 7.13(*a*)-(*c*) the graphs f of this map for three different parameter values corresponding to the trivial fixed point being the only fixed point available (*a*), and to the nontrivial fixed point p being stable (*b*) or unstable (*c*), are depicted. The graphs of the second iterate $f^{(2)}$ for the same parameter values as in (*b*) and (*c*) are drawn in Fig. 7.13(*b'*),(*c'*). We see that when p is a stable fixed point of f, the map $f^{(2)}$ possesses a single nontrivial fixed point. In contrast, when p is unstable for map f, the map $f^{(2)}$ possesses three nontrivial fixed points. One of them is identical to p. According to Section 7.2 the other two (p_1, p_2) define the stable period two orbit of f.

Let $\hat{p}$ be the pre-image of p in the sense that $f(\hat{p}) = p$. We fix our attention on the portion of the graph of $f^{(2)}$ in the interval $[\hat{p}, p]$, which for clarity is enclosed in a box in Figs. 7.13(*b'*),(*c'*). The following observations are worth making:

The interval $[\hat{p}, p]$ is invariant under the action of $f^{(2)}$, just like the unit interval $[0, 1]$ is invariant under the action of f.

Despite marked differences between $f^{(2)}$ and f (number of extrema etc), $f^{(2)}$ restricted to the interval $[\hat{p}, p]$ resembles the graph of the original quadratic map (for a different μ value). Indeed, inside the box it has one fixed point at the end point of this interval (just like f at $x = 0$) and a unique critical point within this interval (just like f at $x = 1/2$).

As μ increases the hump of this quadratic-like map grows until a second, nontrivial fixed point arises within $[\hat{p}, p]$.

In summary, the behavior of $f^{(2)}$ on $[\hat{p}, p]$ is very similar to that of f on $[0,1]$. As μ increases one may therefore expect that the new fixed point born in $[\hat{p}, p]$ for $f^{(2)}$ will become, in turn, unstable and undergo period-doubling (just like p did for f), producing a period four orbit of f. Continuing the procedure outlined in Fig. 7.13 we may likewise find a small box in which the graph of $f^{(4)}$ (the second iterate of $f^{(2)}$) will resemble the original quadratic function, and so on. This suggests that these functions are converging toward a *universal function*. If so, the latter should be accessible through a *renormalization* calculation, expressing that the action of the transformation connecting the original function f to its image in the part of the graph of $f^{(2)}$ inside the box eventually converges to some fixed point. We shall discuss this problem further in the next section.

Fig. 7.13 Illustration of the idea of self-similarity: (a)–(c) the graphs of $f(x)$, eq. (7.13), for $\mu = 0.2$, $\mu = 0.7$ and $\mu = 0.82$ respectively; (b')–(c') the graphs of $f^{(2)}(x)$ for values of μ corresponding to (b) and (c) respectively. $p, \hat{p}$ denote the fixed point and its pre-image. Notice the similarity of the behavior of $f^{(2)}$ in $[\hat{p}, p]$ to that of f in $[0, 1]$.

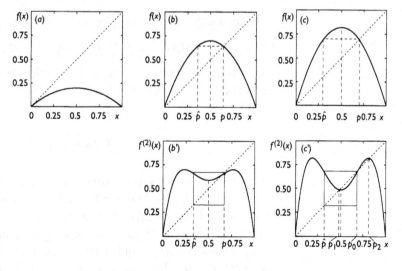

7.5 Routes to chaos: quantitative formulation

We now turn to the part of chaos theory that has attracted so far the greatest attention among all other chaos-related subjects and is largely responsible for the popularity of this theory in the scientific community. Our aim is not to present an exhaustive analysis, which can be found in the abundant original literature and in many excellent monographs, see especially Schuster (1988), Devanay (1989) and Hao (1989). Instead, we shall devote this section to a succinct compilation of the main ideas and turn for the remainder of the chapter to some other aspects of chaos theory which, although very important, are less covered in the literature.

We begin once again with the logistic map, eq. (7.13). The arguments developed in Section 7.4 in connection with Fig. 7.13 and the idea of self-similarity, suggested that f undergoes a series of period doublings as μ increases. In order to formulate this problem in a quantitative manner we first construct a linear mapping taking the fixed point p to 0 and its left pre-image $\hat{p}$ to 1 (actually $\hat{p}$ exists and is on the left of p as long as $\mu > 1/2$):

$$L_\mu(x) = \frac{1}{\hat{p}_\mu - p_\mu}(x - p_\mu) \tag{7.20}$$

where from now on we index all quantities of interest by μ in order to follow clearly the dependence on the parameter. Notice that L_μ expands $[\hat{p}_\mu, p_\mu]$ onto $[0, 1]$ with a change of orientation. Its inverse L_μ^{-1}, obtained by solving (7.20) for x is

$$L_\mu^{-1}(x) = p_\mu + (\hat{p}_\mu - p_\mu)x \tag{7.21}$$

We now define the action of the *renormalization operator* R on $f_\mu(x)$ by

$$Rf_\mu(x) = L_\mu[f_\mu^{(2)}(L_\mu^{-1}(x))] \qquad 0 \le x \le 1 \tag{7.22}$$

One can check by straightforward algebra that $Rf_\mu(x)$ shares many of the properties of f_μ, in particular:

$$\left.\begin{aligned} Rf_\mu(0) &= Rf_\mu(1) = 0 \\ [Rf_\mu(x)]'_{x=1/2} &= 0 \end{aligned}\right\} \tag{7.23}$$

Furthermore, as we saw in connection with Fig. 7.13, Rf_μ converts periodic orbits of period two for f_μ into fixed points for Rf_μ. As long as Rf_μ admits a fixed point $p_1(\mu)$ in the negative slope region (Fig. 7.14) one may identify its pre-image $\hat{p}_1(\mu)$, introduce once again a linear map taking $p_1(\mu)$ to 0 and $\hat{p}_1(\mu)$ to 1 and define a second renormalization. Hence we get another period doubling bifurcation. Continuing this process leads to a succession of period doubling bifurcations as μ increases, as illustrated in Fig. 7.15. It is now conventional (Feigenbaum, 1978) quantitatively to characterize the complexity of this bifurcation diagram by the following two parameters:

The ratio

$$\delta_n = \frac{\mu_n - \mu_{n-1}}{\mu_{n+1} - \mu_n} \tag{7.24a}$$

Numerical simulations show that as $n \to \infty$ δ_n tends to a constant value $\delta = 4.669201 \ldots$ entailing that μ_n converges geometrically to some limit $\mu_\infty < 1$,

$$\mu_n = \mu_\infty - \text{const}/\delta^n \tag{7.24b}$$

The behavior at μ_∞ turns out to be aperiodic and attracting, the attracting set being a fractal.

The separation d_n of the two closest points of a periodic orbit at specific parameter values $\{\tilde{\mu}_n\}$, typically those for which the critical point $x_c = 1/2$

Fig. 7.14 The graphs of f_μ and Rf_μ for the logistic map, eq. (7.13) and for (*a*) $\mu = 0.82$, and (*b*) $\mu = 0.89$.

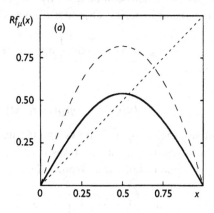

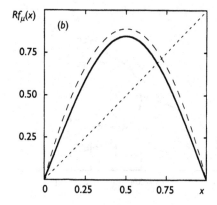

is part of the orbit (one speaks then of *superstable* orbits). Numerical simulations show that the ratio d_n/d_{n+1} also approaches a limit as $n \to \infty$,

$$\lim_{n \to \infty} \frac{d_n}{d_{n+1}} = \alpha \qquad (7.25)$$

which numerically is found to be $\alpha = 2.502907\ldots$.

A most exciting aspect of this *period doubling route to chaos* is Feigenbaum's fundamental observation (Feigenbaum, 1978) that as the above limits are approached, the equivalence of f in $[0,1]$ and $f^{(2)}$ in $[\hat{p}_\mu, p_\mu]$ amounts to the existence of a unique real-valued mapping g such that

$$g(x) = Rg(x) = -\alpha g[g(-x/\alpha)] \qquad (7.26)$$

where α is a scaling factor (related to $(\hat{p}_\mu - p_\mu)^{-1}$ in (7.22) and we have switched to new coordinates in which the origin is placed on the fixed point p_μ. The problem of determining g and α amounts, therefore, to finding the fixed points of the renormalization operator R in the space of real-valued functions. This can be done by successive approximations in which g is expanded in series. One first observes that for any real v,

$$vg(x/v) = -\alpha vg\left[\frac{1}{v}vg(-x/v\alpha)\right]$$

i.e.

$$g(x/v) = -\alpha g[g(-x/v\alpha)]$$

This scale-invariance property allows one to choose $g(0) = 1$. In a first approximation, restricting the solutions to the space of even functions, $g(x) = 1 + bx^2$, one obtains then from (7.26)

$$1 + bx^2 = -\alpha\left[1 + b\left(1 + b\frac{x^2}{\alpha^2}\right)^2\right] = -\alpha\left(1 + b + 2b^2\frac{x^2}{\alpha^2}\right) + O(x^4)$$

Fig. 7.15 The bifurcation diagram for f_μ, showing the successive period doublings at μ_1 (period two), μ_2 (period four), μ_3 (period eight) etc.

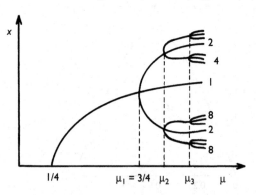

By identifying the coefficients of equal powers of x one finds $b \simeq -1.366$, $\alpha \approx 2.73$, a result quite close to the 'exact' value of α given above. The agreement can be improved by including higher order terms. Rigorous proof of the existence of the fixed point can be found in Lanford (1982) and Campanino and Epstein (1981). The reader will have realized by now that the whole argument is not limited to the specific form of the logistic map but concerns, instead, the entire family of unimodal maps with a nondegenerate critical point satisfying the condition of eq. (7.23). In this respect, therefore, one is entitled to speak of *universality*.

Once the existence of a fixed point of R is established the question naturally arises, how is this fixed point attained by successive iterations. Since R converts periodic orbits of period 2^n for f_μ into periodic orbits of period 2^{n-1} for Rf_μ, we may express the action of the transformation as

$$f_{\mu, n-1} = Rf_{\mu, n} = -\alpha f_{\mu, n}\left[f_{\mu, n}\left(-\frac{x}{\alpha}\right)\right] \tag{7.27}$$

In the spirit of the previous chapters of this book, the natural way to formulate this recurrence problem is to expand this general equation defining the action of R around the fixed point $g(x)$ and truncate to linear terms. Now, we notice that eq. (7.27) actually runs backward in the 'time' variable n. The fixed point is therefore unstable, and repeated use of the linearized renormalization transformation L will be dominated by the eigenvalues of this operator whose absolute value is greater than one (cf. Section 7.2). In other words only expanding directions around the fixed point in the space of functions on which (7.27) is defined will be relevant. Actually the fixed point turns out to be a saddle point. The basic reason for this is that to achieve the fixed point g it is necessary to rescale the control parameter μ at each successive step. This defines a 'critical line' playing the role of the stable manifold of the fixed point. If one is not on this line one can come close to g only up to some finite n. Rescaling beyond such an n will drive the corresponding f_n away from g, a property implying that g behaves as a saddle point. A very similar situation arises in the modern theory of equilibrium critical phenomena (Ma, 1976).

Applied to the period doubling transformation the above procedure produces a dominant eigenvalue $\delta = 4.669201\ldots$, thus providing the fundamental explanation of the geometric law of convergence of μ_n to μ_∞ (eq. (7.24)). But the beauty of renormalization ideas as applied to chaos is that, when appropriately adapted, they can also be used in a variety of problems in which chaos sets in by other mechanisms than the period-doubling cascade. A rather straightforward illustration is provided by intermittency, where one can derive by renormalization arguments the average time spent by the trajectory near the bissectrix (Hu and Rudnick, 1982). A more elaborate, very important extension leading again to universal scaling laws refers to the onset of chaos in the circle map, Section 7.3B (Feigenbaum, Kadanoff and Shenker, 1982).

The quantitative features underlying the various routes to chaos, such as the scaling laws (7.24) and (7.25) have been amply confirmed by experiments in such different areas as chemistry, fluid mechanics or laser physics. This justifies, *a posteriori*, the approach to chaos based on the study of Poincaré recurrences and suggests a remarkable *universality* underlying the onset of chaos in large classes of natural phenomena. A survey of the literature on the quantitative comparison between theory and experiment in this area can be found in Schuster (1988).

7.6 Fully developed chaos: probabilistic description

We now turn our attention to the case where, by one of the mechanisms discussed in the previous sections, chaotic behavior has set in. We have already encountered a number of concrete illustrations of this behavior, such as the logistic map for the value $\mu = 1$ of control parameter (eq. (7.13) and Fig. 7.8) or homoclinic chaos in the Rössler model and in the Belousov–Zhabotinski reaction (eqs. (4.41) and Figs. 4.10, 7.4 and 7.5). So far we do not dispose of quantitative criteria to characterize this phenomenon in comparison to other types of aperiodic behavior like, say, quasi-periodic oscillations. Our objective is to arrive at a sharper view of chaos and to identify some of its key properties. The basis of this program will be the probabilistic approach, to which the present section is devoted.

A number of reasons for undertaking a probabilistic description have already been alluded to in Section 3.4. The crux of the argument was that in nature the process of measurement, by which the observer communicates with a physical system, is limited by a finite precision. As a result, a 'state' of a system is in reality to be understood not as a point in phase space but, rather, as a small region whose size ε reflects the finite precision of the measuring apparatus. Additional sources of delocalization of a dynamical system in phase space also exist in connection, for instance, with incomplete specification of initial data or numerical roundoffs.

If the dynamics of the underlying system were simple the difference between the point-like description and the delocalized description described above would not really matter. The situation changes entirely in the region of chaotic behavior. To illustrate this we depict in Fig. 7.16 the time evolution of two nearby initial data, whose separation ε is supposed to account for the various sources of imprecision or error, for the logistic map at the value $\mu = 1$ of the control parameter. We observe that after an initial stage of coherent coevolution the two curves deviate and eventually their difference becomes comparable to the size of the attractor itself. In other words, chaos amplifies small errors. Put differently, experimentally indistinguishable initial states will eventually evolve to states that are far apart in phase space. To the observer this behavior will signal the inability to predict the future beyond a certain time on the basis of the knowledge of the present conditions. This property, which we also refer to as *sensitivity to initial conditions* and to which we shall return in Section 7.7, introduces a fundamental difference between localized, point-like and delocalized descriptions. It constitutes, therefore, an additional compelling motivation for undertaking a probabilistic description which is the only one to reflect this delocalization and to cope in a natural fashion with irregular, unpredictable successions of events.

Consider a one-dimensional recurrent dynamical system, eq. (7.6). Within the framework of a probabilistic description, the central quantity to evaluate is the probability density $\rho_n(x)$ to be in state x at time n. Let x_0 be an initial state. A point-like description of our dynamical system amounts to stipulating that x_0 is known with an infinite precision. The corresponding probability density is, then $\rho_0(x) = \delta(x - x_0)$. Since after one time unit x_0 is sent to $f(x_0)$, $\rho_0(x)$ will obviously evolve to $\rho_1 = \delta[x - f(x_0)]$.

Suppose now that the initial density $\rho_0(x)$ is a smooth function of x, in the spirit of the ensemble theory point of view discussed in Section 3.4. Obviously, the probability density after one iteration will be given by a superposition of the above point-like evolutions over all initial states x_0 represented in the ensemble. We may therefore write

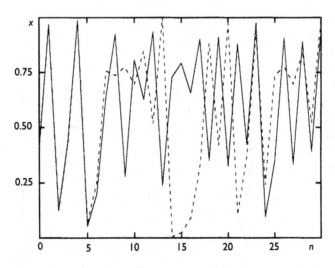

Fig. 7.16 The sensitivity to the initial conditions in the logistic map, eq. (7.13), for $\mu = 1$. Full and dashed lines denote the trajectories emanating from two initial conditions separated by $\varepsilon = 10^{-3}$.

$$\rho_{n+1}(x) = U\rho_n(x) = \int_\Gamma dx_0 \delta[x - f(x_0)]\rho_n(x_0) \qquad (7.28)$$

This equation, known as the Frobenius–Perron equation (Lasota and Mackey, 1985), is the analog of the continuous time evolution equation (3.15) for discrete time dynamical systems. In order to write it in a more explicit form one has to perform the integration over the phase space Γ, in other words, determine the roots x_α of the equation $f(x_\alpha) = x$. The phase space points x_α satisfying this relation are referred to as the *pre-images* of x, since they evolve to x upon one iteration. In a typical one-dimensional endomorphism there are more than one pre-images to a given point x as one sees clearly in Fig. 7.17. Using well-known properties of the delta function one may thus write (7.28) in the equivalent form

$$\rho_{n+1}(x) = \sum_\alpha \frac{1}{|f'(f_\alpha^{-1}(x))|} \rho_n(f_\alpha^{-1}(x)) \qquad (7.29)$$

where f_α^{-1} stands for the branch of the inverse map leading to the particular pre-image x_α. One can check from (7.28) and (7.29) that the Frobenius – Perron operator U preserves the positivity and the normalization of ρ. Such operators are referred to as *Markov operators*.

The Frobenius – Perron equation can be used as the natural starting point of a most illuminating classification of dynamical systems, based on the properties of the invariant probability density $\rho_s(x)$ (the stationary solution of (7.28) and (7.29)) and on the ways time-dependent densities may approach ρ_s in time. We summarize below some salient features and illustrate them on the various examples of chaotic mappings given earlier in this chapter.

Fig. 7.17 The construction leading to the explicit form, eq. (7.29), of the Frobenius–Perron equation for the logistic map, eq. (7.13). $x_{\alpha 1}$, $x_{\alpha 2}$ denote the two pre-images of a state represented by point x.

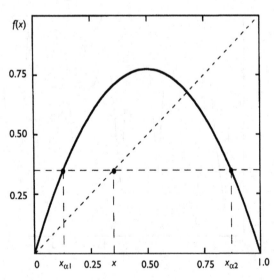

A Ergodicity

According to the Frobenius – Perron equation, the invariant probability density $\rho_s(x)$ satisfies the equation

$$\rho_s(x) = \int_\Gamma dx_0 \delta[x - f(x_0)]\rho_s(x_0) \qquad (7.30)$$

It induces on the subsets C of [0,1] an invariant measure given by

$$\mu_s(C) = \int_C dx\rho_s(x) \qquad (7.31)$$

in the sense that $\mu_s(f^{-1}C) = \mu_s(C)$, where the precise meaning of $f^{-1}C$ is the set of all points x whose images will be in C after one iteration.

In general, under the effect of the dynamics a set C will be transformed into a new one. If this is not the case, the set will be called invariant. We express this property by

$$f^{-1}C = C \qquad (7.32)$$

The dynamical system defined by the iteration law $f(x)$ will be called *ergodic* if every invariant set in the above sense is either the trivial set or the entire interval, in the sense $\mu_s(C) = 0$ or 1. In other words, the phase space of an ergodic system cannot be decomposed into invariant measurable subsets (other than the two previous ones) within each of which the trajectories remain trapped during the evolution.

It is useful to compare ergodicity, also known in the early literature as *metric transitivity*, to a property of chaotic motion of a more geometric nature, namely that an open set in the interval will eventually move under the iteration to cover the entire interval. Ergodicity adds to the above property of *topological transitivity* an important metric element, allowing one to distinguish between 'typical' and 'untypical' trajectories.

A more direct relationship between ergodicity and $\rho_s(x)$ is established by the following theorem (Lasota and Mackey, 1985):

If the dynamical system $f(x)$ is ergodic, then there is exactly one stationary probability density $\rho_s(x)$ which is Lebesgue integrable, in the sense that the integral $\int_C dx\rho_s(x)$ is finite. Furthermore, if there is a unique stationary density $\rho_s(x)$ and $\rho_s(x) > 0$ almost everywhere, then f is ergodic.

Ergodicity is thus intimately connected with the uniqueness and smoothness properties of the invariant probability density. Here and throughout the present section 'almost everywhere' is to be understood as a property being fulfilled in all points of phase space except for a subset of measure zero, the measure to be used being the invariant measure (eq.

(7.31)). Notice that, at this stage, there is nothing which excludes the possibility of singular invariant densities in the form say of delta functions. This is what happens, for instance, in systems possessing a unique fixed point attractor.

Operationally, the invariant density is usually constructed by counting the number of times different regions of phase space are visited by a typical trajectory over a long time interval N, $N \to \infty$. This is how one proceeds, for instance, to compute numerically probability histograms from trajectories. We express this idea by the relation (Eckmann and Ruelle, 1985):

$$\bar{\rho}(x) = \lim_{N \to \infty} \frac{1}{N} \sum_{i=0}^{N-1} \delta[x - f^{(i)}(x_0)] \tag{7.33}$$

In a similar vein one may argue that the outcome of an experimental observation (or numerical simulation) of a certain property $A(x)$ of a dynamical system is also related to the time average of its instantaneous values over a long period,

$$\bar{A}(x) = \lim_{N \to \infty} \frac{1}{N} \sum_{i=0}^{N-1} A[f^{(i)}(x_0)] \tag{7.34}$$

The question therefore naturally arises of how $\bar{\rho}(x)$ and $\bar{A}(x)$ are related to $\rho_s(x)$. The answer is given by the extended *Birkhoff ergodic theorem*:

In an ergodic dynamical system satisfying the condition $\mu_s[f^{-1}(C)] = \mu_s(C)$ for all sets in the unit interval $[0,1]$ the following equalities hold in almost all points in phase space:

$$\bar{A}(x) = \int_\Gamma dx A(x)\rho_s(x) = \langle A(x) \rangle \tag{7.35a}$$

(equality of time and ensemble averages), and

$$\bar{\rho}(x) = \rho_s(x) \tag{7.35b}$$

(equivalence between the 'physical' probability density and the (unique) time-independent Lebesgue-integrable solution of the Frobenius–Perron equation).

As a byproduct, eqs. (7.35) guarantee that the time averages (7.33) and (7.34) are independent of the initial state x_0 for almost all x_0, in the sense defined above.

Let us illustrate the concept of ergodicity on some simple examples. We begin by the twist map, eq. (7.18). Eq. (7.30) reads

$$\rho_s(\theta) = \int_0^1 d\theta_0 \delta[\theta - (\theta_0 + \alpha)]\rho_s(\theta_0)$$

or

$$\rho_s(\theta) = \rho_s(\theta - \alpha) \tag{7.36a}$$

Let α be rational, $\alpha = p/q$, p and q being integers. Eq. (7.36a) then admits a continuous family of normalized solutions

$$\rho_s(\theta) = \frac{1}{q} \sum_{i=0}^{q-1} \delta\left[\theta - \left(\theta^* + \frac{p}{q}i\right)\right] \tag{7.36b}$$

where $\theta^* + (p/q)i$, $i = 0, \ldots, q-1$ are the points of the circle visited by a period-q orbit starting form θ^*. Since this holds for any θ^*, we conclude that there is an infinity of (singular) invariant probability densities supported on disjoint sets, violating the uniqueness condition of the previously enunciated theorem: the system is not ergodic in the full phase space $[0, 1]$. On the other hand it becomes ergodic in the restricted phase space consisting of the points of a periodic orbit emanating from a certain initial ϕ^*. In contrast, if α is irrational, then the validity of (7.36) for all θs entails that there is a unique, smooth solution $\rho_s = \text{const}$ or, taking normalization into account,

$$\rho_s = 1 \qquad \text{(twist map)} \tag{7.37}$$

The system is now ergodic in the full phase space $[0, 1]$. Notice that in the spirit of eq. (7.36b), this solution can be viewed as the unique combination of an infinity of singular measures centered this time on the irrational numbers in the continuum $[0, 1]$.

As a second example consider the tent map $T(x)$, eq. (7.16b). Eq. (7.30) becomes

$$\rho_s(x) = \int_0^{1/2} dx_0 \delta(x - 2x_0)\rho_s(x_0) + \int_{1/2}^1 dx_0 \delta[x - (2 - 2x_0)]\rho_s(x_0)$$

Performing the appropriate change of variables one finally arrives at

$$\rho_s(x) = \frac{1}{2}\left[\rho_s\left(\frac{x}{2}\right) + \rho_s\left(1 - \frac{x}{2}\right)\right] \tag{7.38}$$

which admits the unique smooth, Lebesgue integrable properly normalized invariant density

$$\rho_s(x) = 1 \qquad \text{(tent map)} \tag{7.39}$$

entailing that the invariant measure of a set A for this system is, simply, the length of the corresponding interval (Lebesgue measure). One can deduce from this result the invariant probability of the logistic map, eq. (7.13), for

$\mu = 1$. Indeed, using eq. (7.16a) and the conservation of probability we write

$$\underset{\text{logistic}}{\rho_s(x)\mathrm{d}x} = \underset{\text{tent}}{\rho_s(y)\mathrm{d}y}$$

or, with $\rho_s(y) = 1$,

$$\rho_s(x) = \frac{\mathrm{d}}{\mathrm{d}x}\frac{2}{\pi}\mathrm{arc}\sin(x^{1/2})$$

or finally

$$\rho_s(x) = \frac{1}{\pi[x(1-x)]^{1/2}} \qquad \text{(logistic map)} \qquad (7.40)$$

A common property of all the ergodic invariant densities (7.37), (7.38) and (7.40) is to be delocalized in phase space, despite the purely deterministic origin of the underlying dynamics.† As a result fluctuations around averages are expected to be comparable to the averages themselves. This is to be contrasted with the behavior of the probability densities of thermodynamic fluctuations in systems of interacting particles, which are sharply peaked around well-defined most probable states, except in the immediate vicinity of critical points of phase transitions (Stanley, 1971). In this respect one may therefore view a chaotic dynamical system like the tent map as being permanently in a 'critical' state.

B Mixing and exactness

Despite its interest, the concept of ergodicity deals essentially with static properties such as the existence of an invariant density. Furthermore, as we saw in the previous subsection it cannot discriminate between dynamical systems as different as the twist map with α irrational and the tent map of slope 2. New concepts, addressing dynamical properties, are clearly needed.

One key quantity providing information on the dynamics is the *time autocorrelation function*,

$$C_t(\tau) = \frac{1}{\langle \delta A^2(t) \rangle} \langle \delta A(t+\tau)\delta A(t) \rangle \qquad (7.41)$$

where the average may run over the invariant probability density ρ_s or over the time variable t (cf. eq. (7.35)) and $\delta A = A - \langle A \rangle$, A being a

---

†In addition to such smooth, delocalized densities both systems (logistic and tent) admit an infinity of singular, localized ones supported by the points of the unstable periodic orbits. This does not compromise ergodicity, since these points are of measure zero with respect to the smooth invariant densities (7.39) or (7.40).

certain observable. Similarly, one can define the cross-correlation between two observables A and B or analogous quantities involving measures of sets C in $[0, 1]$. Of particular interest in this respect, is the quantity $\mu_s(C_1 \cap f^{-n}C_2)$, providing the measure of the set of points x belonging to C_1 *and* to the set of points which will evolve into C_2 after n time units. The dynamical system described by $f(x)$ will be called *mixing* if

$$\lim_{n \to \infty} \mu_s(C_1 \cap f^{-n}C_2) = \mu_s(C_1)\mu_s(C_2) \qquad (7.42a)$$

for all sets in the interval $[0,1]$. In other words, the fraction of points starting in C_1 and ending in C_2 after a large number of iterations is just the product of measures of C_1 and C_2, independent of their position in Γ. It is easy to see that any mixing transformation must be ergodic.

There is a very useful connection between mixing and the properties of the time-dependent probability densities, establishing (Lasota and Mackey, 1985) that $f(x)$ is mixing if and only if $U^n \rho$ is weakly convergent to ρ_s for all ρs in the set of probability density functions, in the sense

$$\lim_{n \to \infty} (U^n \rho, g) = (\rho_s, g) \qquad (7.42b)$$

where, $(f, g) = \int \mathrm{d}x f(x)g(x)$ and g is an L^∞-integrable function. Of more interest in physical applications would be the possibility that $U^n \rho$ converges to ρ_s itself. It is here that the concept of *exactness* becomes crucial.

Exact dynamical systems are defined by the property

$$\lim_{n \to \infty} \mu(f^{(n)}C) = 1 \qquad (7.43a)$$

for all sets C in $[0, 1]$ with $\mu(C) > 0$. In plain terms this means that if we follow the evolution of initial conditions within a set C of nonzero measure, then after a large number of iterations the points will have spread and completely filled the entire phase space. It can be proved that exactness implies mixing.

As before there is a useful connection between exactness and time-dependent densities which establishes (Lasota and Mackey, 1985) that in an exact transformation

$$\lim_{n \to \infty} \| U^n \rho - \rho_s \| = 0 \qquad (7.43b)$$

where the L^1 distance is here defined by

$$\| f - g \| = \int_\Gamma |f(x) - g(x)| \, \mathrm{d}x$$

As in the previous subsection, we illustrate now the concepts of mixing and exactness on some simple examples. Consider first the twist map, eq. (7.18). The time-dependent Frobenius–Perron equation (eq. (7.28)) becomes

$$\rho_{n+1}(\theta) = \rho_n(\theta - \alpha) \qquad (7.44)$$

Obviously a nonuniform initial condition – for instance, $\rho(\theta) = 1/\Delta$ if $\theta_0 \le \theta \le \theta_0 + \Delta$, $\rho(\theta) = 0$ otherwise – will merely perform in the course of time a rigid body rotation around an axis perpendicular to the circle while keeping its initial shape, even if α is irrational (Fig. 7.18). We conclude that the twist map does not drive the system to the uniform distribution (7.37): it is a nonexact transformation. Using eq. (7.42b) one can see that it also lacks the mixing property.

The situation is very different for the tent map, eq. (7.16b). Using the same procedure as in Section 7.6A, eq. (7.38), we can write the full time-dependent Frobenius–Perron equation for this system as

$$\rho_{n+1}(x) = \frac{1}{2}\left[\rho_n\left(\frac{x}{2}\right) + \rho_n\left(1 - \frac{x}{2}\right) \right] \qquad (7.45)$$

This equation can be solved exactly by induction,

$$\rho_n(x) = \frac{1}{2^n} \sum_{j=1}^{2^{n-1}} \left[\rho_0\left(\frac{j-1}{2^{n-1}} + \frac{x}{2^n}\right) + \rho_0\left(\frac{j}{2^{n-1}} - \frac{x}{2^n}\right) \right] \qquad (7.46)$$

where ρ_0 is the initial density. In the limit $n \to \infty$, noticing that x remains confined in $[0, 1]$, one sees that the right hand side approaches the integral of ρ_0 over $[0,1]$, which is equal to unity by normalization and thus identical to the invariant density for this system (eq. (7.39)). We conclude that

$$\lim_{n \to \infty} \rho_n(x) = \rho_s(x) \qquad (7.47)$$

The tent map is thus an exact (and mixing as well) dynamical system. An interesting alternative formulation of the convergence of densities to

Fig. 7.18 Evolution of an initial probability density whose support is the arc spanned by the angle Δ, generated by the twist map (eq. (7.18)). The rigid body-like rotation reflects the nonmixing character of the dynamics.

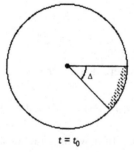

$t = t_0$

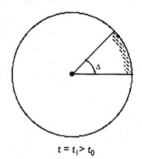

$t = t_1 > t_0$

ρ_s is to consider the functional

$$H_n = \sup_x \rho_n(x) \tag{7.48}$$

where $\sup_x$ designates the supremum of $\rho_n(x)$ in $[0,1]$. We have, using (7.45),

$$H_{n+1} = \frac{1}{2}\sup_x\left[\rho_n\left(\frac{x}{2}\right) + \rho_n\left(1 - \frac{x}{2}\right)\right] \le \frac{1}{2}\left[\sup_x \rho_n\left(\frac{x}{2}\right) + \sup_x \rho_n\left(1 - \frac{x}{2}\right)\right]$$

or finally, noticing that $x/2$ and $1 - x/2$ also run over $[0, 1]$,

$$H_{n+1} \le H_n \tag{7.49}$$

This property is reminiscent of Boltzmann's *H*-theorem, familiar from classical statistical mechanics (Prigogine, 1962). It shows that the tent map irreversibly and monotonously drives the system to a state in which H_n – and thus ρ_n – no longer evolve in time. This state corresponds to the invariant distribution $\rho_s(x) = 1$ constructed in Section 7.6A.

7.7 Error growth, Lyapunov exponents and predictability

We come back now to the property of sensitivity to initial conditions (Fig. 7.16), in the light of the probabilistic formalism developed in Section 7.6. Consider a one-dimensional recurrence in the form of eq. (7.6). Let x_0, $x_0 + \varepsilon$ ($\varepsilon \ll 1$) be two initial conditions slightly displaced with respect to each other. We follow the trajectories emanating from x_0 and $x_0 + \varepsilon$ and evaluate after n time units the instantaneous error

$$u_\varepsilon(n, x_0) = |x_n(x_0 + \varepsilon) - x_n(x_0)| \tag{7.50}$$

or, using eq. (7.6),

$$u_\varepsilon(n, x_0) = |f^{(n)}(x_0 + \varepsilon) - f^{(n)}(x_0)| \tag{7.51}$$

For any given ε, as x_0 runs over the system's attractor the evolution of $u_\varepsilon(n, x_0)$ for finite times is both x_0-dependent and highly irregular. To identify some reproducible trends we therefore perform an average of (7.51) over the attractor. In the spirit of the probabilistic approach this amounts to studying the quantity

$$\langle u_\varepsilon(n)\rangle = \int dx_0 \rho(x_0)|f^{(n)}(x_0 + \varepsilon) - f^{(n)}(x_0)| \tag{7.52}$$

where $\rho(x_0)$ gives the statistical weight of the various points on the attractor. In most cases it appears reasonable to use as the weighting factor the invariant distribution $\rho_s(x_0)$, eq. (7.30).

Fig. 7.19 depicts the evolution of the instantaneous error, eq. (7.52), for the logistic map (eq. (7.13)) averaged over the invariant distribution (7.40). We observe that error growth follows a logistic-like curve (Nicolis and Nicolis, 1991). Three different stages may be distinguished, which actually turn out to be typical of a very wide class of chaotic systems: an initial 'induction' stage during which the error remains small; an intermediate 'explosive' stage displaying an inflection point situated approximately at $n^* \approx \ln 1/\varepsilon$; and a final stage, where the mean error reaches a saturation value u_∞ of the order of the extension of the attractor and remains constant thereafter.

The mechanism ensuring the final saturation of the error is the *reinjection* already mentioned in Section 4.5 in connection with homoclinic chaos, which is also responsible for the fact that the attractor itself remains confined in phase space despite the instability of the motion. In the logistic system this is manifested by the fact that the interval $I_2 = [x_{11}, 1]$ is mapped by the dynamics back into the interval $I_1 = [0, x_{11}]$.

Let us focus on the initial regime. Since the error remains small, an expansion in which only the lowest order term in ε is kept should be meaningful. At the level of eq. (7.51) this leads to

$$u_\varepsilon(n, x_0) = \varepsilon \left| \frac{\mathrm{d}f^{(n)}}{\mathrm{d}x_0} \right| = \varepsilon |f'(x_0)| \dots |f'(x_{n-1})| \qquad (7.53)$$

where $x_1, \dots, x_{n-1}$ are the points visited successively by the 'unperturbed' trajectory emanating from x_0. While these points are to be determined from the full nonlinear equations, the presence of the derivative terms in (7.53) implies that the error itself evolves during this stage according to the

Fig. 7.19 Numerical evaluation of the time dependence of the mean error for the logistic map and for $\mu = 1$. The averaging is performed over 10 000 samples differing in the initial position x_0 which runs over the attractor. The initial error is $\varepsilon = 10^{-4}$.

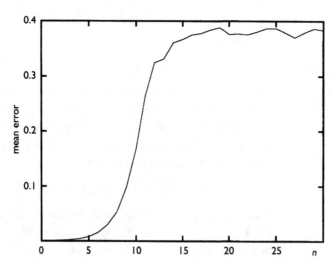

linearized equations around this reference (chaotic) trajectory. These equations define the *tangent space* of the system's attractor (Guckenheimer and Holmes, 1983).

Let us write $u_\varepsilon(n, x_0)$ as

$$u_\varepsilon(n, x_0) = \varepsilon\, e^{n\sigma(n, x_0)} \tag{7.54}$$

Substituting in (7.53) one finds

$$\sigma(n, x_0) = \frac{1}{n} \sum_{i=0}^{n-1} \ln |f'(x_i)| \tag{7.55}$$

We know from Fig. 7.19 that for any finite ε, however small, the error is bound to leave the tangent space after a lapse of time of the order of $n^* = \ln 1/\varepsilon$. But in the idealized case where the limit $\varepsilon \to 0$ is taken n^* is pushed (though very slowly) to infinity and one may define from (7.51)–(7.55),

$$\sigma = \lim_{n \to \infty} \lim_{\varepsilon \to 0} \frac{1}{n} \ln \left| \frac{f^{(n)}(x_0 + \varepsilon) - f^{(n)}(x_0)}{\varepsilon} \right| = \lim_{n \to \infty} \frac{1}{n} \sum_{i=0}^{n-1} \ln |f'(x_i)| \tag{7.56}$$

where in the first equality, the limits are to be taken in the indicated order. We call this quantity the *Lyapunov exponent* of the system. For ergodic transformations the time average in (7.56) should be independent of the initial state (except, perhaps, for a set of points of measure zero) and equal to (cf. (7.34) and (7.35))

$$\sigma = \int dx \rho_s(x) \ln |f'(x)| \tag{7.57}$$

Notice that σ depends on the control parameter in a complex fashion, both through $\rho_s(x)$ and through $f'(x)$.

Applied to the tent map (eq. (7.16b)), eqs. (7.56) and (7.57) give

$$\sigma = \ln 2 \qquad \text{(tent map)} \tag{7.58a}$$

where we used the expression $\rho_s(x) = 1$ for this map (eq. (7.39)). For the logistic map at $\mu = 1$ eq. (7.56) is difficult to evaluate straightforwardly since $|f'(x_i)|$ varies locally as the system runs on the attractor. On the other hand, substituting (7.40) in (7.57) one gets again an explicit result

$$\sigma = \int_0^1 dx \frac{1}{\pi[x(1-x)]^{1/2}} \ln|4 - 8x| = \ln 2 \quad \text{(logistic map)} \tag{7.58b}$$

In view of these results one concludes from (7.54) that in the double limit $\varepsilon \to 0$ (to be taken first) and $n \to \infty$ (to be taken next) the initial error ε

doubles after each iteration. More generally, in a system displaying a positive Lyapunov exponent errors increase exponentially with time *in this double limit*, with a rate equal to the Lyapunov exponent. In this sense the existence of a positive Lyapunov exponent is tantamount to sensitivity to initial conditions and may be regarded as one of the main signatures of chaos.

Appealing as it may seem this view, which still dominates much of the literature on chaos, is unfortunately oversimplified. When confronted with the problem of predicting the evolution of a concrete physical system, the observer is led to follow the growth of a (at best) small but *finite* error over a *transient*, usually small period of time. In this context, the quantity of interest is (cf. (7.52)–(7.55))

$$\langle u_\varepsilon(n) \rangle = \varepsilon \int dx_0 \rho_s(x_0) |f'(x_0)| \dots |f'(x_{n-1})| = \varepsilon \int dx_0 \rho_s(x_0) \, e^{n\sigma(n, x_0)}$$

(7.59)

This equation shows that error growth amounts to studying, for *finite* times, the average over the attractor of an exponential function $<e^{n\sigma(n, x_0)}>$ where $\sigma(n, x_0)$ is given by (7.55). To recover for such ns the picture of a Lyapunov exponent-driven exponential amplification of the error one needs to identify (7.59) with the exponential of the long-time limit of $\sigma(n, x_0)$,

$$\varepsilon \, e^{n \, < (\sigma n, x_0) >} = \varepsilon \, e^{n\sigma}$$

(7.60)

In a typical attractor this is not legitimate since $\sigma(n, x_0)$ is x_0-dependent. This property stems from the *fluctuations* of the local Lyapunov exponents (Nicolis, Mayer-Kress and Haubs, 1983; Grassberger, Badii and Politi, 1988; Nese, 1989; Abarbanel, Brown and Kennel, 1991)

$$\sigma(x_0) = \ln |f'(x_0)|$$

(7.61)

For instance, in the logistic map with $\mu = 1$ the variance of these fluctuations is

$$\langle \delta\sigma^2 \rangle = \langle (\sigma - \langle\sigma\rangle)^2 \rangle = \int_0^1 dx \rho_s(x) [\ln|4 - 8x| - \ln 2]^2 = \frac{\pi^2}{12}$$

(7.62)

which is of the same order as $\langle \sigma \rangle = \sigma$ itself, eq. (7.58b). It is only in the exceptional case of uniform attractors, characterized by a constant local rate of divergence of initial conditions, that (7.59) and (7.60) can be identified. The tent map provides a concrete example, but it must be stressed that most real-world attractors do not satisfy this property.

Let us illustrate the transient behavior of the error for the logistic map at $\mu = 1$ (Nicolis and Nicolis, 1993). Fig. 7.20, open circles, describes the error growth curve obtained from direct numerical evaluation of (7.51), averaged over 100 000 samples of different initial conditions and an initial error of 10^{-5}. We observe a significant deviation from an exponential behavior corresponding to the same initial error and a rate equal to the mean Lyapunov exponent ln2 (shaded circles) in agreement with the previous comments on the difference between (7.59) and (7.60). Writing, in analogy to eq. (7.60),

$$\langle u_\varepsilon(n) \rangle = \varepsilon \, e^{n\sigma_{\text{eff}}} \qquad (7.63)$$

we evaluate σ_{eff} from the simulation data. The results, depicted in Fig. 7.21, show that σ_{eff} is actually n-dependent, starting at $n = 1$ with a value significantly larger than ln 2. This entails that error growth is neither driven by the Lyapunov exponent nor follows an exponential law. This property further complicates the problem of prediction of chaotic systems.

It is interesting to realize from Fig. 7.21 that for ε as small as 10^{-9}, the system never attains a regime where σ_{eff} becomes identical to $\langle \sigma \rangle$. The reason for this has already become clear in the discussion made in connection with Fig. 7.19: at a time $n^* \approx \ln 1/\varepsilon$ the error dynamics leaves the tangent space and evolves toward its saturation value. In the setting of Fig. 7.21 this would give n^* values equal to about 7, 12 and 21 for $\varepsilon = 10^{-3}, 10^{-5}$ and 10^{-9} respectively. These are very short times indeed for the value of infinite time averages such as $\langle \sigma \rangle$ to be established. To keep the error dynamics in the tangent space for, say, 50 time units or so in order possibly to reach a regime driven by $\langle \sigma \rangle$ one would need an initial error of $\varepsilon \approx 10^{-20}$. Such small values hardly ever arise in practical

Fig. 7.20 Short-time behavior of mean error growth with $\varepsilon = 10^{-5}$ obtained from numerical simulation of the logistic map for $\mu = 1$ averaged over 100 000 initial conditions scattered on the attractor (circles); and from the theoretical expression of eq. (7.59) (crosses). The shaded circles correspond to a purely exponential law whose rate is given by $\langle \sigma \rangle = \sigma = \ln 2$.

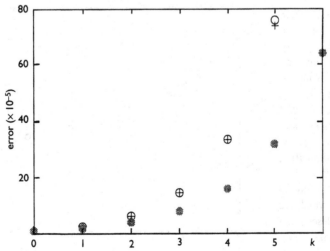

applications, at least at the macroscopic level. In this sense therefore the Lyapunov exponent is to be regarded as one of the quantities which allow one to characterize the attractor (more specifically its tangent space) rather than as the principal quantity monitoring error growth and predictability. In actual fact, as seen from the first equality (7.59), error growth is determined by the *n*-fold time correlation of the system. Such dynamical properties are hard to evaluate analytically, although for the logistic map this can be done for the first few *n*s using symbolic calculus. The results are in full agreement with the conclusions drawn from Fig. 7.20.

We have based the analysis of this section on one-dimensional maps. In more complex – and realistic – systems two further complications are expected to arise. First, a multivariate dynamical system possesses a number of Lyapunov exponents equal to its phase space dimensionality. For short times all of these exponents, including the negative ones associated with motion along the stable manifolds, are expected to take part in the error dynamics. Second, since a typical attractor associated to a chaotic system is fractal, a small error displacing the system from an initial state on the attractor may well place it outside the attractor. Error dynamics will then involve a transient prior to the re-establishment of the attractor, during which errors may well decay in time (Vannitsem and Nicolis, 1994).

We have stressed that deviations of error growth from a Lyapunov exponent-driven exponential law may be generated by the nonuniformity of the attractor. As alluded to already in Section 3.3D, nonuniformity is also at the origin of the inadequacy of the fractal dimension D_0 to describe the structure of such an attractor. Within the framework of the

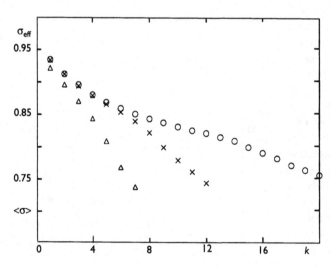

Fig 7.21 Time dependence of the effective rate σ_{eff}, eq. (7.63), obtained numerically as in Fig. 7.20 with $\varepsilon = 10^{-3}$ (triangles); $\varepsilon = 10^{-5}$ (crosses); and $\varepsilon = 10^{-9}$ (circles). For each ε the simulation is carried out till a time $n^* \approx \ln(1/\varepsilon)$ beyond which the linearized description breaks down.

probabilistic description laid down in Section 7.6 one can propose a natural remedy of this deficiency. As in Sections 3.3D and 7.4 one divides the attractor into cells of linear dimension ε, introducing now the additional important element to weight the various points by the probability $P_i(\varepsilon)$ that the trajectory visits cell i. The *generalized dimensions* D_q are then defined in terms of the qth powers of $P_i(\varepsilon)$ through (Renyi, 1970; Hentschel and Procaccia, 1983)

$$D_q = \lim_{\varepsilon \to 0} \frac{1}{q-1} \frac{\ln \sum_{i=0}^{N(\varepsilon)} P_i^q}{\ln \varepsilon} \tag{7.64}$$

where $N(\varepsilon)$ is the total number of cells. For any integer $q > 1$, P_i^q gives the total probability that q points of the attractor are within a given box. This allows one to capture the effect of correlations between different parts of the attractor, and thus the degree of its inhomogeneity.

As expected, for $q = 0$ one recovers from (7.64) the fractal dimension introduced in eq. (3.11). For a general value of q, eq. (7.64) can be handled by methods analogous to those used in the evaluation of the partition function in equilibrium statistical mechanics. This is the starting point of the interesting thermodynamic formalism of fractals (Halsey *et al.*, 1986; Tel, 1987; Bohr and Tel, 1988), the details of which are outside the scope of the present book.

7.8 The dynamics of symbolic sequences: entropy, master equation

We shall now combine the tools of probabilistic analysis laid down in Section 7.6 with the idea of symbolic dynamics introduced in Section 7.4. As we shall see, this blending will provide yet another interesting view of chaos and, in particular, will allow one to sort out some remarkable connections with random processes.

As in Fig. 7.12, we consider a finite partition $C = (C_1, \ldots, C_N)$ of phase space into N nonoverlapping cells. Following eq. (7.31), we also introduce the measure $\mu(C_i)$ of cell i. A natural question to be raised in connection with the developments outlined in the last two sections, is how to quantify the idea that in a chaotic system a localization of the instantaneous state in phase space becomes increasingly difficult as the resolution required gets finer. Now this question is reminiscent of a central problem of information and communication theories (Shannon and Weaver, 1949), namely, what is the amount of data needed to recognize a signal blurred by noise. Assuming that the information source is a random processor possessing N

states, one shows that the central quantity to be studied in this latter context is the *information entropy*

$$H = -\sum_{i=1}^{N} P(i) \ln P(i) \tag{7.65}$$

Indeed, this quantity possesses all the properties compatible with the idea of information needed to localize a system in its state space (Khinchine, 1959):

It takes its largest value for $P(i) = 1/N$, implying that in the case of equiprobable events the amount of data needed to realize one of these events is the maximum one.

The entropy of a composite system CD equals the entropy $H(C)$ of subsystem C plus the (conditional) entropy $H_C(D)$ of subsystem D under the condition that subsystem C is in a given state, a property referred to as *subadditivity*.

Adding an impossible event α, $P(\alpha) = 0$, does not change the entropy of the process.

Notice that information entropy bears an interesting relation with the generalized dimensions D_q, eq. (7.64). Indeed, applying this relation for $q = 1$ one obtains

$$D_1 = \lim_{\varepsilon \to 0} \frac{-\sum_{i=0}^{N(\varepsilon)} P_i(\varepsilon) \ln P_i(\varepsilon)}{\ln (1/\varepsilon)} \tag{7.66}$$

referred to as the *information dimension* of the system. For a given value $D_1 (D_1 \le D_0$ by (7.64)) this allows one to determine how information entropy scales with the size of the boxes in which the fractal object has been subdivided.

Coming back to the phase space partition $C = (C_1, \ldots, C_N)$, by analogy to (7.65) we introduce the information entropy of the partition

$$H(C) = -\sum_{i=1}^{N} \mu(C_i) \ln \mu(C_i) \tag{7.67}$$

In general $\{C_i\}$ are affected by the dynamics. As in earlier sections we denote by $f^{-k}C_i$ the set of points mapped to C_i after k iterations. The set of all $(f^{-k}C_1, \ldots, f^{-k}C_N)$ defines a new partition, denoted by $f^{-k}C$. Continuing the process for n iterations we are led to define a partition $C^{(n)}$ whose cells are the intersection of cells of the partitions C, $f^{-1}C, \ldots, f^{-n+1}C$,

$$C_{i_1} \cap f^{-1}C_{i_2} \cap \cdots \cap f^{-n+1}C_{i_N} \tag{7.68}$$

where $i_1, \ldots, i_N$ run over $1, \ldots, N$. One realizes immediately that as n increases the partition $C^{(n)}$ generated by C becomes finer and finer. Consequently its information entropy $H(C^{(n)})$ will typically be greater than $H(C)$. This can be understood by realizing that with a finer mesh and in the absence of any *a priori* information, more data are needed to localize the system in a particular part of phase space.

The interest of the above construction is that one attains, in the limit $n \to \infty$, an intrinsic property of the dynamics known as the *Kolmogorov–Sinai entropy* h_μ, measuring the average rate of creation of information in time. We first define the rate of creation of information with respect to the partition C:

$$h_\mu(C) = \lim_{n \to \infty} [H(C^{(n+1)}) - H(C^{(n)})] = \lim_{n \to \infty} \frac{1}{n} H(C^{(n)}) \qquad (7.69)$$

In principle this quantity depends on the choice of initial partition C. The Kolmogorov–Sinai (K–S) entropy h_μ is just the supremum of all possible values $h_\mu(C)$, as the partition C gets finer and finer. Notice that for certain partitions C, referred to as *generating partitions*, this latter limit and the supremum may be avoided and one can write $h_\mu(C) = h_\mu$.

Loosely speaking, if h_μ is to be finite $H(C^{(n)})$ should scale as n or, in view of (7.67)–(7.69) the number of cells in the partition $C^{(n)}$ should increase exponentially with n. This is reminiscent of the exponential divergence of initially nearby conditions in the tangent space of a chaotic attractor and suggests that (*a*) the positivity h_μ should be a new way to characterize chaos; and (*b*) there should be a connection between h_μ and Lyapunov exponents. Both statements can, in fact, be justified rigorously. In particular, if the dynamical system possesses a smooth invariant density along the unstable directions of the motion one can establish the *Pesin equality* (Pesin, 1977),

$$h_\mu = \sum_{\sigma_i > 0} \sigma_i \qquad (7.70)$$

where σ_i are the positive Lyapunov exponents.

One can show that the property of $H(C^{(n)})$ to scale linearly with n is also shared by most of the typical random processes like, for instance, Markov processes (Khinchine, 1959). It may thus be regarded as the signature of a process, be it (deterministic) chaos or (random) noise, in which at each new step of the evolution a finite amount of variety is on average created. This implies, in turn, that the memory of the underlying system is short-range. It is legitimate to expect that in many natural systems, particularly in biological systems, this condition is not satisfied. The existence of long-range correlations and memory effects in such systems should be manifested, then, by a sublinear scaling of $H(C^{(n)})$ and thus by a zero K–S entropy (Ebeling and Nicolis, 1992). Such systems may still give rise to quite complex, aperiodic behavior. An interesting notion characterizing this type of complexity is the *topological entropy* (Misiurewicz, 1976), defined for a one-dimensional recurrence $f(x)$ as

$$h_f = \lim_{n \to \infty} \frac{1}{n} \ln N_n \qquad (7.71)$$

where N_n is the number of monotone pieces of $f^{(n)}$.

Let us now follow in a more detailed manner the action of the dynamics at the level of the partition $\{C_1, \dots, C_N\}$. We impose on this partition the condition that each element is mapped by the transformation f on a union of elements (we recall that the cells of the partition are not overlapping). To express this properly we introduce the characteristic function $\chi_C(x)$ of a set C through

$$\chi_C(x) = \begin{cases} 1 & \text{if } x \in C \\ 0 & \text{if } x \notin C \end{cases} \qquad (7.72)$$

The above requirement then amounts to

$$\chi_{f(C_j)} = \sum_{i=1}^{N} a_{ji} \chi_i \qquad j = 1, \dots, N \qquad (7.73)$$

where $\chi_j \equiv \chi_{C_j}$ and the elements of the *topological transition matrix* $\{a_{ji}\}$ take values 0 or 1 depending on whether C_i belongs to $f(C_j)$ or not.

One further condition that we impose is that the state of the system be initially coarse-grained, in the sense that the initial probability density $\rho_0(x)$ is constant within each element of the partition. Writing

$$\rho_0 = \sum_{j=1}^{N} P_0(j) \frac{1}{\mu_j} \chi_j \qquad (7.74a)$$

where for convenience μ_j is chosen to be the Lebesgue measure (rather than the invariant measure, eq. (7.31)) of cell j, we may then identify $P_0(j)$ with the probability of finding the system in cell C_j,

$$P_0(j) = \int_{C_j} \rho_0(x) \, dx \qquad (7.74b)$$

The main problem now is to evaluate the action of the Frobenius–Perron operator on ρ_0. According to eqs. (7.28) and (7.29) we have

$$\rho_1(x) = U\rho_0(x) = \sum_{j=1}^{N} P_0(j) \frac{1}{\mu_j} \sum_{\alpha} \frac{1}{|f'(f_\alpha^{-1}(x))|} \chi_j[f_\alpha^{-1}(x)] \qquad (7.75)$$

We notice that the sum over the branches α of the inverse mapping is nonvanishing only if $x \in f(C_j)$. For any $C_i \subset f(C_j)$ we denote by $f^{-1}_{\alpha(i \to j)}$ those branches of the inverse transformation which map points of C_i into C_j. This means, in particular, that for all $\alpha(i \to j)$ the pre-images of C_i by this branch be contained in C_j,

$$f^{-1}_{\alpha(i \to j)} C_i \subset C_j \tag{7.76}$$

Under these conditions, using eq. (7.73), we can rewrite eq. (7.75) as

$$\rho_1(x) = U\rho_0(x) = \sum_{j=1}^{N} \sum_{i=1}^{N} P_0(j) \frac{1}{\mu_j} a_{ji} \sum_{\alpha(i \to j)} \frac{1}{|f'[f^{-1}_{\alpha(i \to j)}(x)]|} \chi_i(x) \tag{7.77}$$

In contrast to $\rho_0(x)$, the probability density $\rho_1(x)$ is, in general, not coarse-grained, owing to the presence of x-dependences in the coefficients containing the first order derivatives,

$$\frac{1}{|\wedge_{i \to j}|} = \sum_{\alpha(i \to j)} \frac{1}{|f'[f^{-1}_{\alpha(i \to j)}(x)]|} \tag{7.78}$$

A considerable simplification occurs for maps for which these coefficients are x-independent, like in piecewise linear transformations (Grossmann and Thomae, 1977). Indeed, defining

$$P_1(i) = \sum_{j=1}^{N} \frac{\mu_i}{\mu_j} a_{ji} \frac{1}{|\wedge_{i \to j}|} P_0(j) \tag{7.79}$$

one can write eq. (7.77) in a form similar to (7.74a),

$$\rho_1 = \sum_{i=1}^{N} \frac{1}{\mu_i} P_1(i) \chi_i \tag{7.80}$$

entailing that the probability density is again constant within the elements of the partition.

Eq. (7.79) can be conveniently rewritten as

$$\mathbf{P}_1 = \mathbf{W} \cdot \mathbf{P}_0 \tag{7.81}$$

where $\mathbf{P}$ is the one-column matrix col $\{P(1) \dots P(N)\}$ and $\mathbf{W}$ is an $N \times N$ matrix whose elements are given by

$$W_{ij} = \frac{\mu_i}{\mu_j} a_{ji} \frac{1}{|\wedge_{i \to j}|} \tag{7.82}$$

As the Frobenius–Perron operator conserves the norm and the positivity of probability densities the sums of W_{ij} over the rows are equal to unity: $\mathbf{W}$ is a *stochastic matrix*. After n iterations the state of the system is thus entirely describable in terms of probabilities $P_n(j)$ satisfying the equation

(Nicolis and Nicolis, 1988; Nicolis, Piasecki and McKernan, 1992)

$$\mathbf{P}_{n+1} = \mathbf{W} \cdot \mathbf{P}_n \qquad (7.83)$$

The structure of this equation is identical to that of the Chapman–Kolmogorov equations used widely in the theory of stochastic processes (Feller, 1968) which are frequently referred to in the physical literature as *master equations* (Nicolis and Prigogine, 1977; Gardiner, 1983). In other words, under the conditions imposed on the partition (especially eq. (7.73)), which also guarantee the propagation of coarse-graining with time, we have been able to cast the deterministic dynamics into a stochastic process. This mapping gives, at last, a concrete meaning to the statement made repeatedly throughout this book that chaos is associated with a 'random looking' evolution in space and time. There is no contradiction whatsoever between this result and the deterministic origin of chaos: in the probabilistic view we look at our system through a 'window' (phase space cell), whereas in the deterministic view it is understood that we are exactly running on a trajectory. This is, clearly, an unrealistic assumption in view of our earlier comments on finite precision and roundoff errors.

The simplest type of stochastic process satisfying eq. (7.83) are Markovian processes, for which memory extends only one step backward in time. The converse, however, is not true: a non-Markovian process can satisfy the master equation (Feller, 1959; Courbage and Hamdan, 1991).

Let us illustrate the above construction on the example of the tent map, eq. (7.16b). We choose the two-cell partition $C = \{C_1, C_2\}$ where $C_1 = [0, x_{11}]$, $C_2 = [x_{11}, 1]$, $x_{11} = \frac{2}{3}$ being the nontrivial fixed point. This partition obviously satisfies the requirement expressed in eq. (7.73), by the very definition of the fixed point and the fact that the angular point of $f(x)$ at $x = \frac{1}{2}$ is mapped to $x = 1$. More specifically we find

$$f(C_1) = C_1 \cup C_2, \qquad f(C_2) = C_1$$

entailing that the topological transition matrix $\{a_{ji}\}$ is of the form

$$\mathbf{a} = \begin{pmatrix} 1 & 1 \\ 1 & 0 \end{pmatrix} \qquad (7.84)$$

We come now to the construction of the stochastic transition matrix $\mathbf{W}$, eq. (7.82). Referring to Fig. 7.22 we see that a point x_2 in C_2 has two pre-images in C_1. There are thus two inverse branches sending C_2 into C_1, entailing (cf. (7.78)) that

$$\frac{1}{|\Lambda_{2 \to 1}|} = 1 \qquad (7.85a)$$

since $|f'| = 2$ for the tent map. In contrast a point x_1 in C_1 has one pre-image in C_2 and one in C_1, entailing that

$$\frac{1}{|\wedge_{1\to1}|} = \frac{1}{2}, \qquad \frac{1}{|\wedge_{1\to2}|} = \frac{1}{2} \qquad (7.85b)$$

Introducing (7.84) and (7.85) into (7.82) we obtain

$$\mathbf{W} = \begin{pmatrix} \frac{1}{2} & 1 \\ \frac{1}{2} & 0 \end{pmatrix} \qquad (7.86)$$

The eigenvalues λ_i and eigenvectors $\mathbf{u}_i$ of this matrix are

$$\lambda_1 = 1, \qquad \mathbf{u}_1 = \mathrm{col}(\tfrac{2}{3}, \tfrac{1}{3}) \qquad (7.87a)$$

$$\lambda_2 = -\tfrac{1}{2}, \qquad \mathbf{u}_2 = \mathrm{col}(-1, 1) \qquad (7.87b)$$

The first one corresponds, in our coarse-grained description, to the properly normalized invariant state $\mathbf{P}_\infty$ of the system. Notice that its structure is compatible with the invariant distribution $\rho_s(x) = 1$ (eq. (7.39)) obtained earlier from the Frobenius–Perron equation. As for the second one, it is responsible for the relaxation of an initial state, coarse-grained over the two-cell partition, toward this invariant state.

Fig. 7.22 Two-cell partition $\{C_1, C_2\}$ in the tent map, defined by the end points 0, 1 and the fixed point x_{11}. A point $x_1 \in C_1$ has one pre-image $f^{-1}_{1(1\to1)}(x_1)$ in C_1 and one preimage $f^{-1}_{1(1\to2)}(x_1)$ in C_2. A point $x_2 \in C_2$ has two pre-images $f^{-1}_{1(2\to1)}(x_2)$ and $f^{-1}_{2(2\to1)}(x_2)$ in C_1.

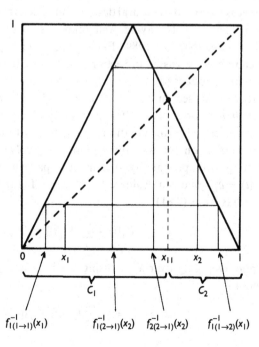

The procedure can easily be extended to more refined partitions. One interesting class is provided by the partitions formed by the end points of the interval and the points on an unstable periodic orbit of the map. For instance, for the three-cell partition of the tent map $\{C_1 = [0, \frac{2}{5}],$ $C_2 = [\frac{2}{5}, \frac{4}{5}], C_3 = [\frac{4}{5}, 1]\}$ where $(\frac{2}{5}, \frac{4}{5})$ are the points of the period two cycle, one obtains

$$
\mathbf{W} = \begin{pmatrix} \dfrac{1}{2} & 0 & 1 \\[2mm] \dfrac{1}{2} & \dfrac{1}{2} & 0 \\[2mm] 0 & \dfrac{1}{2} & 0 \end{pmatrix}
$$

and

$$\lambda_1 = 1, \qquad \mathbf{u}_1 = \mathrm{col}(\tfrac{2}{5}, \tfrac{2}{5}, \tfrac{1}{5}) \qquad \text{(invariant distribution)}$$

$$\lambda_{2,3} = \frac{\pm \mathrm{i}}{2}, \qquad \mathbf{u}_{2,3} = \mathrm{col}(-1 \circlearrowleft \mathrm{i}, \pm \mathrm{i}, 1)$$

This structure turns out to be general: as the number of cells of the partition is increased by taking cycles of increasing order, the eigenvalues (other than the invariant one) are determined by the nth roots of (-1).

One of the applications of the master equation approach described above is to provide a natural description of deterministic chaos in terms of *strings of symbols* having well-defined statistical properties (Nicolis, Nicolis and Nicolis, 1989; Ebeling and Nicolis, 1992). This view has already been discussed in Section 7.4. Among the various quantities that one can introduce to characterize such sequences are the entropies defined earlier in this section (eqs. (7.65), (7.67), (7.69)) which give a measure of their *information content*. The possibility of casting the dynamics into a discrete closed-form equation provides additional insight, as it gives information on the way these quantities evolve in time. For instance, starting with probability one from a particular cell of the partition $P_0(i) = \delta_{i\alpha}^{kr}$, one can evaluate the change of entropy after one iteration. One has from (7.81)

$$P_1(i) = \sum_j W_{ij} \delta_{j\alpha}^{kr} = W_{i\alpha}$$

Upon averaging over all αs using the invariant distribution $P_\infty(\alpha)$ one obtains

$$\Delta S = S_1 = -\sum_{\alpha j} P_\infty(\alpha) W_{j\alpha} \ln W_{j\alpha} > 0 \qquad (7.88)$$

which is nothing but the analog of the Kolmogorov–Sinai entropy in our master equation description.

A master equation of the form (7.83) is also characterized by an irreversible approach to a unique invariant distribution $P_\infty(i)$, such that

$$\mathbf{P}_\infty = \mathbf{W} \cdot P_\infty \qquad (7.89)$$

To see this we introduce the quantity

$$H_n = \sum_j P_\infty(j) F\left(\frac{P_n(j)}{P_\infty(j)}\right) \qquad (7.90a)$$

where $F(x)$ is a convex function of its argument. A familiar choice of such a function is $F(x) = x \ln x$, in which case (7.90a) reduces to the relative entropy

$$H_n = \sum_j P_n(j) \ln \frac{P_n(j)}{P_\infty(j)} \qquad (7.90b)$$

We now consider the time evolution of H_n. Using eq. (7.83) we have

$$H_{n+1} - H_n = \sum_j P_\infty(j) \left\{ F\left[\sum_k \frac{P_\infty(k)}{P_\infty(j)} W_{jk} \frac{P_n(k)}{P_\infty(k)} \right] - F\left[\frac{1}{P_\infty(j)} P_n(j) \right] \right\} \qquad (7.91)$$

At this point we notice that, by (7.89),

$$\sum_k \frac{P_\infty(k)}{P_\infty(j)} W_{jk} = 1$$

Invoking the property of convex functions

$$F\left(\sum_k m_k x_k\right) \leq \sum_k m_k F(x_k), \quad \sum_k m_k = 1 \qquad (7.92)$$

we may thus transform (7.91) into

$$H_{n+1} - H_n \leq \sum_j P_\infty(j) \left\{ \sum_k \frac{P_\infty(k)}{P_\infty(j)} W_{jk} F\left[\frac{P_n(k)}{P_\infty(k)} \right] - F\left[\frac{P_n(j)}{P_\infty(j)} \right] \right\}$$

or, using the properties of the stochastic matrix W_{jk},

$$H_{n+1} - H_n \leq 0 \qquad (7.93)$$

where by (7.92) the equality sign applies only when $\mathbf{P}_n = \mathbf{P}_\infty$. This is reminiscent of the *H*-theorem of statistical mechanics (Kac, 1959). It shows that at the level of the probabilistic description the unpredictability of deterministic chaos beyond the Lyapunov time is replaced by full predictability, in the sense that the statistical state of the system evolves in a regular manner. This suggests the interesting possibility of *statistical forecasting* of complex dynamical systems giving rise to chaos.

Last but not least, the master equation approach to deterministic chaos provides a systematic approximation scheme for computing the eigenvalues and eigenfunctions of the Frobenius – Perron operator itself, whose spectral properties still remain largely unknown (McKernan and Nicolis, 1994). A full technical presentation of this complex subject is outside the scope of the present book. For some interesting alternative recent studies we refer the reader to Gaspard (1992a,b), Hasegawa and Saphir (1992), and Antoniou and Tasaki (1993) among others. We close this section by pointing out that there is also an interesting connection between the eigenvalues of the Frobenius–Perron operator and the decay properties of the time correlation functions, introduced in Section 7.6 in connection with the property of mixing. These are, in turn, related to the singularities of the power spectrum – the Fourier transform of the autocorrelation function – in the complex frequency plane (Ruelle 1985,1986). For certain types of system (typically those with uniform attractors) it can be shown that the only such singularities are poles, entailing that the autocorrelation function decays exponentially in time. The general case remains, however, open.

7.9 Spatio-temporal chaos

We have seen, in Chapters 1 and 6 of this book, that large classes of spatially extended systems may undergo a sequence of transitions leading to regimes displaying aperiodic dependence in both space and time, which we referred to rather loosely as *spatio-temporal chaos*. We already know from Section 6.8 that certain normal form equations such as the complex Landau–Ginzburg equation lead to such a regime beyond the instability of the uniform limit-cycle solution, and that defects seem to play an important role in its onset. From Chapters 1, 2 and Section 6.9 it is also clear that many other equations encountered in physical sciences, such as the Navier–Stokes equation, the reaction – diffusion equations and the Kuramoto–Shivashinski equation may also generate spatio-temporal chaos. The objective of this final section of the present chapter is to have a new look at this phenomenon in the light of the tools and ideas of chaos theory developed in the previous sections.

Just as temporal chaos differs from other forms of time-dependent behavior by the coexistence of a large number of interacting time scales, spatio-temporal chaos will likewise be associated with the property of displaying a large number of interacting space scales. As it happens this is also one of the signatures of fully developed turbulence, a ubiquitous feature of large-scale flows. This analogy constitutes one of the main motivations for studying spatio-temporal chaos.

As noticed already in Section 3.1, a spatially extended system possesses, in principle, an infinite number of variables. We have seen already that in some cases truncation to a finite small number can legitimately be carried out. By the very definition of spatio-temporal chaos given above, such truncations should break down in the presence of this regime. In other words, in studying spatio-temporal chaos one has to cope at the outset with the additional complexity arising from the presence of an infinite number of degrees of freedom. Three key questions arise under these new circumstances:

Can one identify well-defined scenarios leading to spatio-temporal chaos?
Is it still possible to characterize the attractor of an infinite-dimensional dynamical system giving rise to spatio-temporal chaos?
How (if at all) is sensitivity to initial conditions manifested in such a system?

In the following we deal with these questions and give some concrete illustrations of the phenomenology of spatio-temporal chaos.

A Some scenarios leading to spatio-temporal chaos

In contrast to the situation described in Sections 7.3–7.5, the transition to spatio-temporal chaos has not yet been characterized by global quantitative laws comparable to, say, the ones governing the period doubling cascade. Still, experience acquired through linear stability analysis, numerical simulation and laboratory experiments suggests a number of interesting features that should undoubtedly be parts of the more comprehensive theory yet to come.

The main point relates to some ideas already developed in Section 6.5: in systems of large spatial extent the size L becomes a natural control parameter. By increasing L from small values while keeping the other, more traditional, control parameters fixed the threshold of a first instability can be reached. As L is increased further, the number of unstable modes grows (in the Kuramoto–Shivashinsky equation it does so linearly with L) and new instability thresholds are encountered. Frequently these thresholds lead first to periodic, multiperiodic or weakly chaotic behavior in time; next to space symmetry-breaking that may include the appearance of defects; and eventually to spatio-temporal chaos. The latter can have, in turn, some structure in space in the form of irregular patches of spatial activity immersed in 'laminar' well-organized domains referred to as spatio-temporal intermittency (Kaneko, 1989;

Chaté, 1989; Daviaud *et al.*, 1990, 1992); or a completely incoherent, fully 'turbulent' type of spatial activity.

When a second control parameter is varied in addition to L various instability thresholds can again be reached, even at fixed L. A typical signature of large systems is, then, that the larger the L the closer these instability thresholds will lie relative to each other.

An interesting question is how to characterize the complexity of the observed pattern. A number of interesting ideas have been advanced such as spatial and spatio-temporal Fourier spectra; symbolic encoding of a spatial pattern by introducing a partition in physical space expressing the values of the state variables relative to pre-assigned thresholds; or statistical quantities like entropies, the histogram of 'laminar' regions as a function of their size, life-time distributions, and so forth. Details of these attempts, still in a largely exploratory stage, can be found in the above mentioned references.

B The attractor of systems giving rise to spatio-temporal chaos

It has been stressed repeatedly in this book that spatially distributed systems possess an infinite number of degrees of freedom. We have seen that when these systems operate in the close vicinity of the transition to a regular spatial pattern, say the Rayleigh–Bénard instability, a drastic reduction can take place. The relevant variables then obey a normal form equation and lie on a low-dimensional attractor. The question we raise here is: what is the nature of the attractor when the system operates in the regime of spatio-temporal chaos?

In Chapter 3 we have introduced the concept of the *universal attractor*, the largest bounded invariant set of phase space toward which all trajectories of the dynamical system converge as time tends to infinity. We know already that the structure of such an attractor may be quite complicated even in systems with a small number of variables, since the attractor may be a fractal set. A remarkable recent result already briefly mentioned in Section 6.5, is that in many spatially distributed systems, although the phase space is an infinitely-dimensional function space, the universal attractor has *finite* fractal dimension (Constantin and Foias, 1985; Constantin, Foias and Temam, 1985). Still it may be quite complicated and attract the trajectories very slowly. For this reason it is desirable to embed the attractor in an *inertial manifold* (Constantin *et al.*, 1989): a finite-dimensional invariant manifold Σ toward which the solutions tend with at least a uniform exponential rate, in the sense that the distance

$$\text{dist}\,(\phi_t(\mathbf{X}_0), \bar{\Sigma}) \le c\,e^{-kt} \tag{7.94}$$

for $t \ge t_0 > 0$, c being a positive constant.

The flow restricted to the inertial manifold is equivalent to that of a system with a finite number of degrees of freedom. We refer to the evolution laws in such a phase space as an *inertial form*. An inertial form constitutes thus an optimal finite-dimensional representation of the original system as far as long-term dynamics – and hence bifurcation analysis – is concerned.

The results surveyed above are, essentially, existence theorems. They leave open two questions of crucial practical importance: (*a*) can the dimensionality D of the universal attractor be estimated; and (*b*) can the inertial form be inferred from the original evolution laws?

An appropriate way to formulate (*a*) is to ask how D scales with the size parameter L, it being understood that in the limit $L \to \infty$, D will also tend to infinity. Typical estimates give the power law behavior

$$D < L^a \tag{7.95}$$

with a greater than unity – for instance, $a = 13/8$ for the Kuramoto–Shivashinsky equation (Nicolaenko, Scheurer and Temam, 1985). For large size systems of interest this leads to large numbers of little practical relevance. There is evidence (Manneville, 1985; Pomeau, Pumir and Pelcé, 1984) that in some cases D may actually scale linearly with L.

We come briefly now to question (*b*). Consider the standard form of the evolution laws of a spatially distributed system, eqs. (6.52). We assume that the linear operator $\mathscr{L}$ has a spectrum consisting of negative eigenvalues $\lambda_1 \ge \lambda_2 \ge \ldots$ and a complete orthonormal set of eigenfunctions. In terms of these eigenfunctions one can then construct an operator $\mathbf{P}_m$ projecting $\mathbf{X}$ from the original phase space Γ into the subspace $\gamma = \mathbf{P}_m\Gamma$ spanned by the first m eigenfunctions, and its orthogonal complement $\mathbf{Q}_m = \mathbf{1} - \mathbf{P}_m$. Operating with $\mathbf{P}_m$ on both sides of (6.52) one gets

$$d\mathbf{p}/dt = \mathscr{L}\cdot\mathbf{p} + \mathbf{P}_m\cdot\mathbf{h}(\mathbf{p} + \mathbf{q}) \tag{7.96}$$

where $\mathbf{p} \in \mathbf{P}_m\Gamma$, $\mathbf{q} \in \mathbf{Q}_m\Gamma$. This equation is not closed, as it must be supplemented with an equation for $\mathbf{q}$. In most approaches the inertial manifold is represented as a mapping $\mathbf{q} = \phi(\mathbf{p})$ from the space $\mathbf{P}_m\Gamma$ into the space $\mathbf{Q}_m\Gamma$. The inertial form then reads

$$d\mathbf{p}/dt = \mathscr{L}\cdot\mathbf{p} + \mathbf{P}_m\cdot\mathbf{h}[\mathbf{p} + \phi(\mathbf{p})] \tag{7.97}$$

Usually the mapping $\phi(\mathbf{p})$ can only be determined in an approximate manner. For a recent survey we refer the reader to Brown, Jolly, Kevrekidis and Titi (1990).

C Sensitivity to initial conditions in spatio-temporal chaos

As mentioned in the end of Section 7.7, a multivariate dynamical system possesses a number of Lyapunov exponents equal to its phase space dimensionality. In a spatially distributed system this will result in an infinite number of exponents. If the conditions of existence of a finite-dimensional universal attractor and inertial manifold discussed previously are met, the number of positive Lyapunov exponents will remain finite, although possibly very large, for any given value of the size parameter L. But in the limit in which $L \to \infty$ one expects that, similarly to the attractor dimension D (eq. (7.95)), this number will tend to infinity. Since the largest exponent is likely to remain bounded, this will result in a *continuous spectrum* of positive Lyapunov exponents. The question therefore naturally arises, of whether in the presence of chaos this new feature will leave a typical signature in the way sensitivity to initial conditions will be manifested.

In Section 7.7 we saw that, already in one-dimensional chaotic mappings, sensitivity to initial conditions may be quite intricate. In the limit of uniform attractors characterized by the same local rate of divergence in each point of phase space it reduces, however, to an exponential law whose exponent is given by the (unique) positive Lyapunov exponent. In order to disentangle the role of the continuous spectrum from that of the nonuniformity of the attractor we therefore focus on the case of spatio-temporal chaos displaying a constant local rate of divergence.

Since there is no known example of a continuous time system in which such a property holds true, we turn to a useful alternative model of spatio-temporal complexity provided by *coupled map lattices* (Kaneko, 1989). Specifically, we consider a one-dimensional lattice of diffusively coupled cells

$$x_{n+1}(j) = f[x_n(j)] + \frac{D}{2}[x_n(j+1) + x_n(j-1) - 2x_n(j)]$$

$$j = 1, \ldots, N \tag{7.98}$$

where n is a discrete time, j the lattice point, x a continuous variable and $f(x)$ a function describing the local dynamics, typical examples of which have been given in Section 7.3.

Let $\bar{x}_n(j)$ be a reference state solution of (7.98) corresponding to spatio-temporal chaos. The tangent space of the attractor of the system in this regime is defined by (see Section 7.7)

$$\xi_{n+1}(j) = f'(\bar{x}_n(j))\xi_n(j) + \frac{D}{2}[\xi_n(j+1) + \xi_n(j-1) - 2\xi_n(j)] \tag{7.99}$$

We notice that if $f'(\bar{x}_n(j))$ were a constant λ_0, eq. (7.99) could be solved exactly. On the other hand, this is precisely the case of constant local rate of divergence for which we have been looking. The advantage of coupled map lattices is to allow one to realize this possibility. An example is provided by the choice

$$f(x) = \begin{cases} 2x & 0 \le x \le \frac{1}{2} \\ 2x - 1 & \frac{1}{2} \le x \le 1 \end{cases} \tag{7.100}$$

This dynamical system, referred to as the *Bernoulli map*, is quite similar to the tent map (eq. (7.16b)). When used in eq. (7.98) it generates, for appropriate choices of D, spatio-temporal chaos as illustrated in Fig. 7.23. To see what happens in the tangent space we expand $\xi_n(j)$ in a Fourier series,

$$\xi_n(j) = \sum_{k=0}^{N-1} \tilde{\xi}_n(k)\, e^{i(2\pi/N)jk} \tag{7.101}$$

where we have assumed periodic boundary conditions. Substituting into (7.99) we find the following equations for the mode amplitudes $\tilde{\xi}_n(k)$

$$\tilde{\xi}_{n+1}(k) = \left\{ \lambda_0 + D\left[\cos\left(\frac{2\pi}{N}k\right) - 1 \right] \right\} \tilde{\xi}_n(k) \tag{7.102}$$

(notice that $\lambda_0 = 2$ for the Bernoulli map). The solution of this equation is strictly exponential,

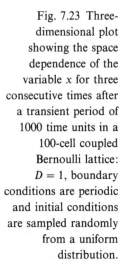

Fig. 7.23 Three-dimensional plot showing the space dependence of the variable x for three consecutive times after a transient period of 1000 time units in a 100-cell coupled Bernoulli lattice: $D = 1$, boundary conditions are periodic and initial conditions are sampled randomly from a uniform distribution.

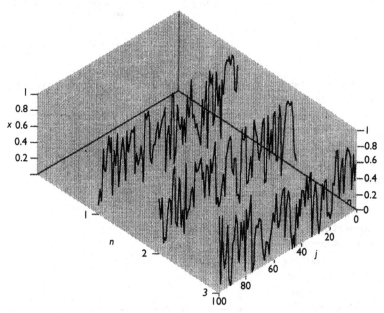

$$\tilde{\xi}_n(k) = a_k \, e^{n\sigma_k} \qquad\qquad (7.103a)$$

with

$$\sigma_k = \sigma_0 + \ln\left\{1 + \frac{D}{\lambda_0}\left[\cos\left(\frac{2\pi}{N}k\right) - 1\right]\right\} \qquad k = 0, \ldots, N - 1$$

$$(7.103b)$$

where $\sigma_0 = \ln\lambda_0$ is the maximum Lyapunov exponent. The set of $\{\sigma_k\}$ provides the full spectrum of the Lyapunov exponents. We notice that in the limit $N \to \infty$ the spectrum becomes continuous. The contributions of the modes k such that $k \ll N$ can then be approximated by expanding $\cos[(2\pi/N)k]$, yielding

$$\sigma_k \approx \sigma_0 + \ln\left\{1 - \frac{2D\pi^2}{\lambda_0 N^2}k^2\right\} \qquad\qquad (7.104)$$

In the context of sensitivity to initial conditions we must see how $\xi_n(j)$ behaves in time. To this end one substitutes (7.103a) into (7.101) after determining a_k from the initial conditions. Choosing for simplicity initial perturbations around the reference state $\bar{x}_n(j)$ localized on the box j_0,

$$\xi_0(j) = \varepsilon \delta_{j, j_0}^{kr} \qquad\qquad (7.105)$$

and using (7.104) one finds after some algebra

$$\xi_n(j_0) \approx \varepsilon \, e^{n\sigma_0} \frac{1}{N} \sum_{k=0}^{N-1} \cos\left[\frac{2\pi}{N}(j - j_0)k\right] e^{-n(2D/\lambda_0)(\pi^2 k^2/N^2)}$$

$$\approx \varepsilon \, e^{n\sigma_0} \int_{-\infty}^{\infty} dx \, e^{-n(2D/\lambda_0)\pi^2 x^2} \approx \frac{\varepsilon}{(2\pi D/\lambda_0)^{1/2}} e^{n\sigma_0} \frac{1}{\sqrt{n}} \qquad (7.106)$$

where the validity of the asymptotic evaluation of the sum requires

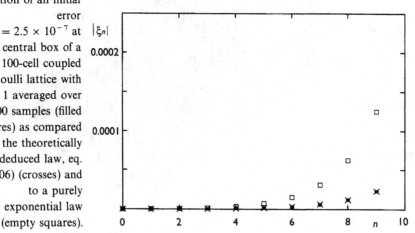

Fig. 7.24 Time evolution of an initial error $|\xi_0| = 2.5 \times 10^{-7}$ at the central box of a 100-cell coupled Bernoulli lattice with $D = 1$ averaged over 1000 samples (filled squares) as compared to the theoretically deduced law, eq. (7.106) (crosses) and to a purely exponential law (empty squares).

$n > (2\pi^2 D/\lambda_0)^{-1}$.

Eq. (7.106) shows that the accumulation of Lyapunov exponents around the maximum value σ_0 arising in the limit of a size parameter $N \to \infty$, introduces a modulation of the purely exponential growth $e^{n\sigma_0}$ of an initial error ε in the form of an inverse power law (Nicolis, Nicolis and Wang, 1992). Fig. 7.24 summarizes the result of a numerical simulation carried out on the full nonlinear system, along with a comparison with the analytic result of eq. (7.106). The system, composed of $N = 100$ cells, is first run with random initial conditions for a sufficient amount of time to reach its attractor. After this transient period a small localized perturbation is imposed on the central cell and the trajectories of the perturbed as well as the unperturbed systems in each box are subsequently monitored. The filled squares of the figure depict the evolution of the absolute value of instantaneous error in the central box as deduced from the above two trajectories. Crosses indicate the analytic result of eq. (7.106), and empty squares a purely exponential amplification with a rate equal to the maximum Lyapunov exponent σ_0. We see that the numerical and analytic results are practically indistinguishable for short times for which the linearized regime is expected to be valid, whereas, on the contrary, appreciable deviations from a purely exponential law show up at a rather early stage.

As it turns out this conclusion applies to a whole class of spatially distributed systems generating spatio-temporal chaos. This suggests that an inverse power law modulation of exponential error growth can be regarded as a characteristic signature, at the level of sensitivity to initial conditions, of spatially distributed systems of large extension.

Problems

7.1 Using the x-axis as the Poincaré 'surface' of section construct the explicit form of the Poincaré map induced by the normal form equations for the Hopf bifurcation, eqs. (5.55). Check that the fixed point of the recurrence corresponds to the limit cycle solution of the flow and that the stability condition (7.10b) is verified for a supercritical Hopf bifurcation.

7.2 The Schwarzian derivative of a function f at x is defined as

$$Sf(x) = \frac{f'''(x)}{f'(x)} - \frac{3}{2}\left(\frac{f''(x)}{f'(x)}\right)^2$$

Show that for the logistic map $Sf(x) < 0$ for all x. (The relevance of this property stems from the theorem (Devanay, 1989) that for unimodal maps, $Sf(x) < 0$ implies that there exists at most one stable periodic orbit for each parameter value.)

7.3 Show that the logistic map at $\mu = 0.959\,75$ possesses a stable period-three orbit ($x_1 = 0.149\,888$, $x_2 = 0.489\,172$, $x_3 = 0.959\,299$). By the theorem mentioned in Problem 7.2 this is the only stable solution available for this value of μ. On the other hand, a remarkable theorem by Sarkovskii (Devanay, 1989) establishes a hierarchical relation between orbits of different periods by showing that if a map possesses a periodic orbit of period k it automatically possesses an orbit of period l provided $k \rhd l$, with:

$$3 \rhd 5 \rhd 7 \rhd \ldots 2 \cdot 3 \rhd 2 \cdot 5 \rhd \ldots 2^2 \cdot 3 \rhd 2^2 \cdot 5 \rhd \ldots 2^k 3 \rhd 2^k 5 \ldots$$
$$\rhd 2^k \rhd \ldots 2^2 \rhd 2^1 \rhd 1.$$

It follows, therefore, that for $\mu = 0.959\,75$ the logistic map possesses unstable periodic orbits of all the periods of the above list. Notice that if a unimodal map has only finitely many periodic orbits, then their periods are necessarily powers of two.

7.4 Consider the Bernoulli map

$$B(x) = \begin{cases} 2x & 0 \le x \le \frac{1}{2} \\ 2x - 1 & \frac{1}{2} < x \le 1 \end{cases}$$

and its analog with negative slope

$$\bar{B}(x) = \begin{cases} 1 - 2x & 0 \le x \le \frac{1}{2} \\ 2 - 2x & \frac{1}{2} < x \le 1 \end{cases}$$

Sketch the graphs of these maps and determine the number and stability of their periodic orbits of period n.

7.5 A qualitative description of instabilities leading to chaotic dynamics in a variety of chemical systems and in mathematical models of chaos such as Rössler's model (eqs. (4.41)) can be obtained from a one-dimensional map possessing a maximum followed by a minimum. The simplest realization of this structure is the cubic map (Fraser and Kapral, 1982; Kapral and Fraser, 1984)

$$f(x, a, b) = ax^3 + (1 - a)x + b(1 - x^2), \qquad a > 0$$

Determine the fixed points and the period two solutions of this dynamical system and study their stability in the two parameter space (a,b).

7.6 Using eq. (7.30) for the invariant density ρ_s prove that $\mu_s(C) = \mu_s(f^{-1}C)$. What kind of regularity properties should one impose on f for this relation to make sense?

7.7 Using the definitions of Section 7.7 construct the analytic form of the probability distribution of the local Lyapunov exponent $\sigma(x) = \ln |f'(x)|$ for the tent map and for the logistic map at $\mu = 1$. Calculate the first two moments of this distribution and compare with (7.58) and (7.62).

7.8 Consider a piecewise linear map consisting of three segments

$$f(x) = \begin{cases} a_1 x + b_1 & 0 \le x \le x_1 \\ a_2 x + b_2 & x_1 < x \le x_2 \\ a_3 x + b_3 & x_2 < x \le 1 \end{cases}$$

with $a_i > 0$, $a_1 < a_2 < a_3$ (such maps qualitatively describe the dynamics of homoclinic systems referred to in Section 4.5, see Arnéodo *et al.* (1993).

(a) Determine the conditions on the parameters such that each of the three intervals of definition of $f(x)$ is mapped by the dynamics into a union of intervals.
(b) Under these conditions compute the invariant density and the Lyapunov exponent and check that the first stages of error growth are superexponential, in agreement with Section 7.7.

7.9 Consider the tent map

$$f(x) = \begin{cases} rx & 0 \le x \le \frac{1}{2} \\ r(1-x) & \frac{1}{2} < x \le 1 \end{cases}$$

(a) Prove that for $r = (\sqrt{5} + 1)/2$ the Frobenius–Perron equation induces a master equation on the partition.

$$\left\{ C_1 = \left[r\left(1 - \frac{r}{2}\right), \frac{1}{2} \right], \quad C_2 = \left[\frac{1}{2}, \frac{r}{2} \right] \right\}$$

Compute the invariant density and show that the system is mixing.
(b) Prove that for $r = \sqrt{2}$ a coarse-grained description in terms of two cells (which ones?) is still possible, but that the system now loses its mixing property while still possessing a nonsingular invariant density (Nicolis, Piasecki and McKernan, 1992).

7.10 Consider three diffusively coupled logistic maps (eq. (7.98)). Perform linear stability analysis of the homogeneous fixed point and of the homogeneous period-two orbit in the domain of parameters in which each of them is stable toward homogeneous perturbations. Can a diffusion-induced instability occur in this system?

Proof of the principle of linearized stability for one-variable systems

The evolution equation in the original variable X reads (cf. eqs. (3.6))

$$\mathrm{d}X/\mathrm{d}t = F(X) \tag{A1.1}$$

where F is a scalar function of the single (scalar) variable X. Introducing the decomposition of eq. (3.20) into a reference steady-state solution X_s and a perturbation x,

$$X = X_s + x \tag{A1.2}$$

we obtain (cf. (3.25) and (3.26))

$$\mathrm{d}x/\mathrm{d}t = F'_s x + \tfrac{1}{2}F''_s x^2 + \tfrac{1}{6}F'''_s x^3 + \cdots$$

$$\equiv \lambda x + a_2 x^2 + a_3 x^3 + \cdots \tag{A1.3}$$

Here $F_s^{(k)}$ denotes the kth derivative of F with respect to X evaluated at X_s. For convenience we choose F'_s to be the control parameter figuring in the original eqs. (3.6).

The 'auxiliary' linearized system associated to (A1.3) is

$$\mathrm{d}x/\mathrm{d}t = \lambda x \tag{A1.4a}$$

and admits the solution

$$x = x_0 \, \mathrm{e}^{\lambda t} \tag{A1.4b}$$

which is obviously asymptotically stable for $\lambda < 0$, unstable for $\lambda > 0$ and Lyapunov stable for $\lambda = 0$.

To see how the full nonlinear equation behaves under these conditions we multiply both sides by x, transforming (A1.3) into

$$\tfrac{1}{2}\mathrm{d}r/\mathrm{d}t = \lambda x^2 + a_2 x^3 + \cdots$$

$$= g(x) \tag{A1.5}$$

where $r = x^2$ is a nonnegative variable. By construction, the function $g(x)$ on the right hand side of this equation is such that

$$g(0) = 0, \; g'(0) = 0, \; g''(0) = 2\lambda \tag{A1.6}$$

Using Rolle's theorem, familiar from calculus (Sokolnikoff, 1939), we may therefore write

$$g(x) - g(0) = \frac{x^2}{2} g''(\theta x), \; 0 < \theta < 1 \tag{A1.7}$$

and consequently

$$\mathrm{d}r/\mathrm{d}t = rg''(\theta x) \tag{A1.8}$$

We now consider, successively, the cases $\lambda < 0$, $\lambda > 0$ and $\lambda = 0$.

(i) $\lambda < 0$. From (A1.6) this is tantamount to $g''(0) < 0$. By continuity this entails $g''(\theta x) < 0$ in a sufficiently small neighborhood of the origin. It follows from (A1.8) that

$$\mathrm{d}r/\mathrm{d}t < 0 \tag{A1.9}$$

or equivalently that $|x|$ decreases to zero and X_s is thus asymptotically stable, just as predicted by the linearized eq. (A1.4a). The situation can be further illustrated by plotting the original function F versus X (Fig. A1.1(a)). Again, since $\lambda < 0$ $F'_s < 0$. By continuity $F'(X)$ remains negative in a vicinity of X_s. A small positive perturbation leading the system to X_1 will 'see' a negative value of F. By (A1.1) X_1 will then decrease toward X_s. A similar argument can be made concerning the evolution of a negative initial perturbation leading the system to X_2. This proves, once again, the asymptotic stability of X_s in the nonlinear regime.

(ii) $\lambda > 0$. From (A1.6) $g''(0) > 0$ and, by continuity, $g''(\theta x) > 0$ as well. From (A1.8) r and thus $|x|$ increase then in time, thereby establishing the instability of X_s. The situation can again be illustrated on the graph of F (Fig. A1.1(b)). In short we find, again, agreement with the behavior predicted by the linearized eq. (A1.4a).

(iii) $\lambda = 0$. In this intermediate case $g''(0) = 0$. Nothing can be said now about $g''(\theta x)$, since the property of a function to vanish on a particular point is 'nongeneric' in the sense that, typically, it does not extend to a neighborhood of that point. No general statement on stability can thus be made, in other words, the conclusions of the linear analysis based on (A1.4) are not necessarily valid. A more detailed study is needed: this was precisely the assertion made in Section 3.6.

An example of what may happen in the nonlinear range when the linearized problem predicts Lyapunov stability but not asymptotic stability is given in Fig. A1.1(c). We see that an initial positive perturbation will evolve farther and farther from X_s, whereas a negative one will evolve toward X_s. Clearly, the fixed point is no longer Lyapunov stable.

To illustrate further the connection between the stability properties of the original nonlinear system and its linearized version we consider the logistic equation (2.51),

$$dX/dt = kX(1 - X/N) \tag{A1.10}$$

We are interested in the stability properties of the reference state of population extinction, $X_{s1} = 0$. The linearized equation around this state reads

$$dx/dt = kx \tag{A1.11}$$

It is asymptotically stable for $k < 0$, unstable for $k > 0$ and Lyapunov stable for $k = 0$.

On the other hand the full differential equation (A1.10) is separable and can thus be solved exactly. One obtains straightforwardly

$$X(t) = \frac{N}{\dfrac{N - X(0)}{X(0)} e^{-kt} + 1} \tag{A1.12}$$

Fig. A1.1 Illustration of the mechanism behind the principle of linearized stability. (a) $\lambda < 0$, (b) $\lambda > 0$, (c) $\lambda = 0$

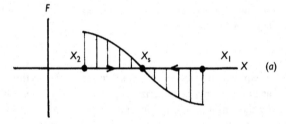

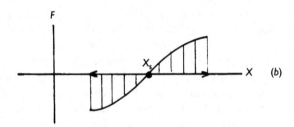

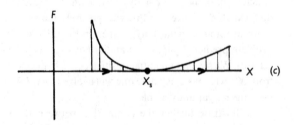

where $X(0)$ is the initial condition. We observe that:

For $k < 0$, $\lim_{t \to \infty} X(t) = 0 : X_{s1} = 0$ is asymptotically stable.
For $k > 0$, $\lim_{t \to \infty} X(t) = N : X_{s1} = 0$ is unstable, and the system reaches the second fixed point solution $X_{s2} = N$ available.

Exceptionally, in this particular problem the agreement with the predictions of the linearized equation extends to the case of Lyapunov stability ($k = 0$) as well.

Hopf bifurcation analysis of the Brusselator model

We use B as the bifurcation parameter. The explicit form of eqs. (5.1) is (cf. eqs. (2.44))

$$\frac{d}{dt}\begin{pmatrix} x \\ y \end{pmatrix} = \begin{pmatrix} B-1 & A^2 \\ -B & -A^2 \end{pmatrix}\begin{pmatrix} x \\ y \end{pmatrix} + \begin{pmatrix} \dfrac{B}{A}x^2 + 2Axy + x^2y \\[2mm] -\dfrac{B}{A}x^2 - 2Axy - x^2y \end{pmatrix}$$

$$\mathscr{L}(B) \qquad\qquad\qquad \mathbf{h}(\mathbf{x}, B) \qquad\qquad\qquad (A2.1)$$

As we saw in Section 4.4, eqs. (4.36)–(4.38), the fixed point $(A, B/A)$ undergoes an instability at

$$B_c = A^2 + 1 \qquad (A2.2a)$$

at which the real part of the eigenvalue of the linearized operator $\mathscr{L}$ vanishes while the imaginary part is

$$\mathrm{Im}\,\omega_c = \Omega_c = A \qquad (A2.2b)$$

To evaluate the solutions in the vicinity of B_c we insert (5.7), (5.8), (5.46) and (5.47) into eqs. (A2.1). We outline hereafter the calculations to the first few significant orders.

A $O(\varepsilon)$

Eq. (5.48) takes the explicit form

$$A\frac{\partial}{\partial T}\begin{pmatrix} x_1 \\ y_1 \end{pmatrix} = \begin{pmatrix} B_c - 1 & A^2 \\ -B_c & -A^2 \end{pmatrix}\begin{pmatrix} x_1 \\ y_1 \end{pmatrix}$$

$$= \begin{pmatrix} A^2 & A^2 \\ -A^2 - 1 & -A^2 \end{pmatrix}\begin{pmatrix} x_1 \\ y_1 \end{pmatrix} \qquad (A2.3)$$

The critical eigenvector $\mathbf{u}$ associated to (A2.3) is given by

$$\begin{pmatrix} A^2 & A^2 \\ -(A^2+1) & -A^2 \end{pmatrix} \begin{pmatrix} u_1 \\ u_2 \end{pmatrix} = iA \begin{pmatrix} u_1 \\ u_2 \end{pmatrix}$$

or, taking the first line of this vector equation,

$$(A^2 - iA)u_1 + A^2 u_2 = 0 \qquad (A2.4)$$

Since normalization of **u** is arbitrary we choose to write it as

$$\begin{pmatrix} 1 \\ \dfrac{u_2}{u_1} \end{pmatrix}$$

or, using (A2.4),

$$\mathbf{u} = \begin{pmatrix} 1 \\ \dfrac{u_2}{u_1} \end{pmatrix} = \begin{pmatrix} 1 \\ \dfrac{i-A}{A} \end{pmatrix} \qquad (A2.5)$$

The solution of eq. (A2.3) is thus

$$\begin{pmatrix} x_1 \\ y_1 \end{pmatrix} = c(\tau_1, \tau_2) \begin{pmatrix} 1 \\ \dfrac{i-A}{A} \end{pmatrix} e^{iT} + cc \qquad (A2.6)$$

The adjoint $\mathscr{L}^+(B_c)$ of $\mathscr{L}(B_c)$ and its critical eigenvector $\mathbf{u}^+$ (associated to the eigenvalue $-iA$) are given by

$$\mathscr{L}^+(B_c) = \begin{pmatrix} A^2 & -(A^2+1) \\ A^2 & -A^2 \end{pmatrix} \qquad (A2.7a)$$

and

$$\mathbf{u}^+ = \begin{pmatrix} 1 \\ \dfrac{u_2^+}{u_1^+} \end{pmatrix} = \begin{pmatrix} 1 \\ \dfrac{A^2 + iA}{A^2 + 1} \end{pmatrix} \qquad (A2.7b)$$

Notice that $\mathbf{u}^+ e^{iT}$ is the null eigenvector of the operator $\mathbf{J}_T^+$ defined in (5.51b).

B $O(\varepsilon^2)$

From eqs. (5.11), (5.50) and (A2.1) we get

$$\left[A \frac{\partial}{\partial T} \begin{pmatrix} 1 & 0 \\ 0 & 1 \end{pmatrix} - \begin{pmatrix} A^2 & A^2 \\ -(A^2+1) & -A^2 \end{pmatrix} \right] \begin{pmatrix} x_2 \\ y_2 \end{pmatrix}$$

$$= \gamma_1 \begin{pmatrix} 1 & 0 \\ -1 & 0 \end{pmatrix} \begin{pmatrix} x_1 \\ y_1 \end{pmatrix} + \begin{pmatrix} \dfrac{A^2 + 1}{A} x_1^2 + 2Ax_1 y_1 \\ -\dfrac{A^2 + 1}{A} x_1^2 - 2Ax_1 y_1 \end{pmatrix}$$

$$- \frac{\partial}{\partial \tau_1} \begin{pmatrix} x_1 \\ y_1 \end{pmatrix} \tag{A2.8}$$

Using (A2.6) and (A2.7b) we write the solvability condition for this inhomogeneous equation as

$$\int_0^{2\pi} dT\, e^{-iT} \left(1, \frac{A^2 - iA}{A^2 + 1}\right) \left\{ \gamma_1 c \begin{pmatrix} 1 \\ -1 \end{pmatrix} e^{iT} - \frac{\partial c}{\partial \tau_1} \begin{pmatrix} 1 \\ \dfrac{i - A}{A} \end{pmatrix} e^{iT} \right.$$

$$+ \left(\begin{matrix} \dfrac{A^2 + 1}{A}(c^2 e^{2iT} + |c|^2) + 2Ac^2 \dfrac{i - A}{A} e^{2iT} + 2A|c|^2 \dfrac{-i - A}{A} \\ -\dfrac{A^2 + 1}{A}(c^2 e^{2iT} + |c|^2) - 2Ac^2 \dfrac{i - A}{A} e^{2iT} - 2A|c|^2 \dfrac{-i - A}{A} \end{matrix} \right) + \text{cc} \bigg\}$$

$$\tag{A2.9}$$

The integration over T cancels the contributions of all parts in the curly bracket except those containing e^{iT}. This yields

$$\partial c / \partial \tau_1 = \tfrac{1}{2} \gamma_1 c \tag{A2.10a}$$

from which it follows that

$$\gamma_1 = 0, \partial c / \partial \tau_1 = 0 \tag{A2.10b}$$

The second order eq. (A2.8) now simplifies to

$$\left[A \frac{\partial}{\partial T} \begin{pmatrix} 1 & 0 \\ 0 & 1 \end{pmatrix} - \begin{pmatrix} A^2 & A^2 \\ -(A^2 + 1) & -A^2 \end{pmatrix} \right] \begin{pmatrix} x_2 \\ y_2 \end{pmatrix}$$

$$= \begin{pmatrix} \dfrac{A^2 + 1}{A} x_1^2 + 2Ax_1 y_1 \\ -\dfrac{A^2 + 1}{A} x_1^2 - 2Ax_1 y_1 \end{pmatrix} \tag{A2.11}$$

The solvability condition (A2.10a) being identically satisfied we proceed to solve this equation by noticing that, on the grounds of (A2.6), the right hand side features contributions either T-independent or depending on T through the factors e^{2iT} or e^{-2iT}. We thus seek a solution of (A2.11) in the form

$$\begin{pmatrix} x_2 \\ y_2 \end{pmatrix} = \begin{pmatrix} p_0 \\ q_0 \end{pmatrix} + \begin{pmatrix} p_2 \\ q_2 \end{pmatrix} e^{2iT} + \text{c.c.} \tag{A2.12}$$

Substituting in (A2.11) one finds, after a straightforward algebra,

$$p_0 = 0$$

$$q_0 = \frac{2}{A^3}(A^2 - 1)|c|^2$$

$$p_2 = \frac{-2i}{3A^2}(1 - A^2 + 2iA)c^2 \qquad \left.\begin{array}{l}\\[3em]\end{array}\right\} \text{(A2.13)}$$

$$q_2 = \frac{1}{3A^3}[1 - 5A^2 + 2iA(2 - A^2)]c^2$$

C $O(\varepsilon^3)$

From eqs. (5.12), (5.50) and (A2.1) we get, keeping in mind that $\gamma_1 = 0$, $\partial c/\partial\tau_1 = 0$:

$$\left[A\frac{\partial}{\partial T}\begin{pmatrix}1 & 0\\ 0 & 1\end{pmatrix} - \begin{pmatrix}A^2 & A^2\\ -(A^2 + 1) & -A^2\end{pmatrix}\right]\begin{pmatrix}x_3\\ y_3\end{pmatrix}$$

$$= \gamma_2\begin{pmatrix}1 & 0\\ -1 & 0\end{pmatrix}\begin{pmatrix}x_1\\ y_1\end{pmatrix} - \frac{\partial}{\partial\tau_2}\begin{pmatrix}x_1\\ y_1\end{pmatrix}$$

$$+ \left[\frac{2(A^2 + 1)}{A}x_1x_2 + 2A(x_1y_2 + x_2y_1) + x_1^2 y_1\right]\begin{pmatrix}1\\ -1\end{pmatrix} \quad \text{(A2.14)}$$

The solvability condition for this inhomogeneous equation can be written out explicitly using (A2.6), (A2.7b), (A2.12) and (A2.13). As in the $O(\varepsilon^2)$ case, only the parts of the right hand side depending on T through the oscillating factor e^{iT} will contribute. This yields

$$\left(1, \frac{A^2 - iA}{A^2 + 1}\right)\cdot\left(\gamma_2 c\begin{pmatrix}1\\ -1\end{pmatrix} - \frac{\partial c}{\partial\tau_2}\begin{pmatrix}1\\ \frac{i - A}{A}\end{pmatrix} + |c|^2 c\right.$$

$$\left\{2\frac{A^2 + 1}{A}\frac{-2i}{3A^2}(1 - A^2 + 2iA) + \frac{4(A^2 - 1)}{A^2} + \frac{2}{3A^2}[1 - 5A^2 + 2iA(2 - A^2)]\right.$$

$$\left.+ \frac{4i}{3A}(1 - A^2 + 2iA)\frac{i + A}{A} - \frac{i + A}{A} + 2\frac{i - A}{A}\right\}\begin{pmatrix}1\\ -1\end{pmatrix}\Big) = 0 \quad \text{(A2.15)}$$

Performing the scalar product and reestablishing the original variables, parameters and time scale we finally arrive at the explicit form of the normal form equation (cf. eq. (5.53))

$$\frac{dz}{dt} = \frac{B - B_c}{2}z - \left(\frac{A^2 + 2}{2A^2} + \frac{i}{2}\frac{4A^4 - 7A^2 + 4}{3A^3}\right)|z|^2 z \quad \text{(A2.16)}$$

Notice that the factor $\frac{1}{2}$ in the linear term of the right hand side is just the derivative of the eigenvalue ω of the linearized operator with respect to B evaluated at the criticality (cf. eq. (4.36)).

References

H. Abarbanel, R. Brown and M. Kennel, 1991. Variation of Lyapunov exponents on a strange attractor, *J. Nonlinear Sci.* **1**, 175–99.

R. Abraham and C. Shaw, 1985. *Dynamics – the geometry of behavior* (Part Two). Aerial Press, Santa Cruz.

G. Ahlers and R. Behringer, 1978. The Rayleigh–Bénard instability and the evolution of turbulence, *Progr. Theor. Phys. Suppl.* **64**, 186–201.

G. Ahlers, P. Hohenberg and M. Lücke, 1984. Externally modulated Rayleigh–Bénard convection: experiment and theory, *Phys. Rev. Lett.* **53**, 48–51.

P. Allen, 1975. Darwinian evolution and a predator – prey ecology, *Bull. Math. Biol.* **37**, 389–405.

Y. Almirantis and G. Nicolis, 1987. Morphogenesis in an asymmetric medium, *Bull. Math. Biol.* **49**, 519–30.

A. Andronov, A. Vitt and C. Khaikin, 1966. *Theory of oscillators*, Pergamon, Oxford.

I. Antoniou and S. Tasaki, 1993. Spectral decomposition of the Renyi map, *J. Phys.* **A26**, 73–94.

F. Argoul, A. Arnéodo, P. Richetti and J.C. Roux, 1987. From quasi-periodicity to chaos in the Belousov–Zhabotinski reaction. I. Experiment *J. Chem. Phys.* **86**, 3325–38.

R. Aris, 1975. *The mathematical theory of reaction and diffusion in porous catalysts*, Oxford University Press, Oxford.

A. Arnéodo and O. Thual, 1988. Approche du chaos par la théorie des formes normales : de la théorie à l'expérience, in *Le Chaos*, P. Bergé ed., Eyrolles, Paris.

A. Arnéodo, F. Argoul, J. Elezgaray and P. Richetti, 1993. Homoclinic chaos in chemical systems, *Physica* **D62**, 134–69.

V. Arnol'd, 1980. *Chapitres supplémentaires de la théorie des équations différentielles ordinaires*, Mir, Moscow.

D. Arrowsmith and C. Place, 1990. *An introduction to dynamical systems*, Cambridge University Press, Cambridge.

A. Babloyantz and A. Destexhe, 1986. Low dimensional chaos in an instance of epileptic seizure, *Proc. Nat. Acad. Sci. USA* **83**, 3513–17

G. Baker and J. Gollub, 1990. *Chaotic dynamics*, Cambridge University Press, Cambridge.

L. Bauer, H. Keller and E. Reiss, 1975. Multiple eigenvalues lead to secondary bifurcation, *SIAM Rev.* **17**, 101–22.

P. Bergé, Y. Pomeau and C. Vidal, 1984. *L'ordre dans le chaos*, Hermann, Paris.

T. Bohr and T. Tel, 1988. The thermodynamics of fractals, in *Directions in chaos*, B.L. Hao ed., World Scientific, Singapore.

H. Brand, 1990. Phase dynamics – the concept and some recent developments, in *Patterns, defects and materials instabilities*, D. Walgraef and N. Ghoniem eds., Kluwer Academic, Dordrecht.

H. Brown, M. Jolly, I. Kevrekidis and E. Titi, 1990. Use of approximate inertial manifolds in bifurcation calculations, in *Continuation and bifurcations: numerical techniques and applications*, D. Roose, B. De Dier and A. Spence eds., Kluwer Academic, Dordrecht.

N. Brown and L. Wolpert, 1990. The development of handedness in left/right asymmetry, *Development* **109**, 1–9.

L. Bunimovitch and Y. Sinai, 1980. Markov partitions for dispersed billiards, *Commun. Math. Phys.* **78**, 247–80.

J. Burgers, 1948. A mathematical model illustrating the theory of turbulence, *Adv. Appl. Mech.* **1**, 171–99.

F. Busse and J. Whitehead, 1971. Instabilities of convection cells in a high Prandtl number fluid, *J. Fluid Mech.* **47**, 305–20.

F. Busse, 1978. Nonlinear properties of thermal convection, *Rep. Progr. Phys.* **41**, 1921–67.

M. Campanino and H. Epstein, 1981. On the existence of Feigenbaum's fixed point, *Comm. Math. Phys.* **79**, 261–302.

D. Campbell, A. Newell, R. Schrieffer and H. Segur (eds), 1986. Solitons and coherent structures, *Physica* **D18**, 1–480.

V. Castets, E. Dulos, J. Boissonade and P. De Kepper, 1990. Experimental evidence of a sustained standing Turing-type nonequilibrium chemical pattern, *Phys. Rev. Lett.* **64**, 2953–6.

L. Cesari, 1963. *Asymptotic behavior and stability problems in ordinary differential equations*, Academic, New York.

S. Chandrasekhar, 1961. *Hydrodynamic and hydromagnetic stability*, Oxford University Press, Oxford.

H. Chaté, 1989. *Transition vers la turbulence via intermittence spatio-temporelle*, Ph.D. dissertation, University Pierre and Marie Curie, Paris.

P. Collet and J.P. Eckmann, 1980. *Iterated maps on the interval as dynamical systems*, Birkhäuser, Basel.

P. Constantin and C. Foias, 1985. Global Lyapunov exponents, Kaplan – Yorke formulas and the dimension of the attractors for 2-D Navier – Stokes equations, *Comm. Pure and Appl. Math.* **38**, 1–27.

P. Constantin, C. Foias and R. Temam, 1985. Attractors representing turbulent flows, *Mem. Amer. Math. Soc.* **53**, no 314.

P. Constantin, C. Foias, B. Nicolaenko and R. Temam, 1989. *Integral manifolds and inertial manifolds for dissipative partial differential equations*, Springer, Berlin.

P. Coullet and E. Spiegel, 1983. Amplitude equations for systems with competing instabilities, *SIAM J. Appl. Math.* **43**, 776–821.

P. Coullet and L. Gil, 1988. Normal form description of broken symmetries, *Solid State Phenomena.* **3–4**, 57–76.

M. Courbage and D. Hamdan, 1991. A class of nonmixing dynamical systems with monotonic semi group property, *Lett. Math. Phys.* **22**, 101–6.

M. Cox, G. Ertl and R. Imbihl, 1985. Spatial self-organization of surface structures during an oscillating catalytic reaction, *Phys. Rev. Lett.* **54**, 1725–8.

C. Crafoord and E. Källén, 1978. A note on the condition for existence of more than one steady-state solution of the Budyko – Sellers type models, *J. Atmos. Sci.* **35**, 1123–5.

F. Daviaud, M. Bonetti and M. Dubois, 1990. Transition to turbulence via spatio-temporal intermittency in one-dimensional Rayleigh–Bénard convection, *Phys. Rev.* **A42**, 3388–99.

F. Daviaud, J. Lega, P. Bergé, P. Coullet and M. Dubois, 1992. Spatio-temporal intermittency in a 1-d convective pattern : theoretical model and experiments, *Physica* **D55**, 287–308.

P. De Boer, 1986. Thermally driven motion of highly viscous fluids, *Int. J. Heat Mass Transfer* **29**, 681–8.

S. De Groot and P. Mazur, 1962. *Nonequilibrium thermodynamics*, North Holland, Amsterdam.

A. Destexhe, 1992. *Aspects non-linéaires de l'activité rythmique du cerveau*, Ph.D. dissertation, University of Brussels.

R. Devanay, 1989. *Chaotic dynamical systems*, Addison-Wesley, Redwood City, CA.

G. Dewel, P. Borckmans and D. Walgraef, 1984. Spatial structures in nonequilibrium systems, in *Chemical instabilities*, G. Nicolis and F. Baras eds., Reidel, Dordrecht.

A. De Wit, 1993. *Brisure de symétrie spatiale et dynamique spatio-temporelle dans les systèmes réaction – diffusion*, Ph.D. dissertation, University of Brussels.

M. Dubois and P. Bergé, 1981. Instabilités de couche limite dans un fluide en convection : évolution vers la turbulence, *J. Phys.* **42**, 167–74.

F. Dumortier, 1991. Local study of planar vector fields : singularities and their unfoldings, in *Structures in dynamics*, H. Broer, F. Dumortier, S. Van Strien and F. Takens eds., North Holland, Amsterdam.

W. Ebeling and G. Nicolis, 1992. Word frequency and entropy of symbolic sequences : a dynamical perspective, *Chaos, Solitons and Fractals* **2**, 635–50.

V. Eckhaus, 1965. *Studies in nonlinear stability theory*, Springer, Berlin.

J.P. Eckmann and D. Ruelle, 1985. Ergodic theory of chaos and strange attractors, *Rev. Mod. Phys.* **57**, 617–56.

M. Eigen and P. Schuster, 1979. *The hypercycle*, Springer, Berlin.

T. Erneux and M. Herschkowitz-Kaufman, 1975. Dissipative structures in two dimensions, *Biophys. Chem.* **3**, 345–54.

G. Ertl, 1991. Oscillatory kinetics and spatio-temporal self-organization in reactions at solid surfaces, *Science*, **254**, 1750–5.

J. Feder, 1988. *Fractals*, Plenum, New York.

M. Feigenbaum, 1978. Quantitative universality for a class of nonlinear transformations, *J. Stat. Phys.* **19**, 25–52.

M. Feigenbaum, L. Kadanoff and S. Shenker, 1982. Quasiperiodicity in dissipative systems : a renormalization group, *Physica* **5D**, 370–86.

D. Feinn, P. Ortoleva, W. Scalf, S. Schmidt and M. Wolff, 1978. Spontaneous pattern formation in precipitating systems, *J. Chem. Phys.* **67**, 27–39.

W. Feller, 1959. Non-markovian processes with the semigroup property, *Ann. Math. Stat.* **30**, 1252–3.

W. Feller, 1968. *An introduction to probability theory and its applications* Vol. 1, 3rd edition, Wiley, New York.

R. Field, E. Körös and R. Noyes, 1972. Oscillations in chemical systems. II. Thorough analysis of temporal oscillation in the bromate–cerium–malonic acid system, *J. Am. Chem. Soc.* **94**, 8649–64.

R. Field, 1975. Limit cycle oscillations in the reversible Oregonator, *J. Chem. Phys.* **63**, 2289–96.

R. Field and M. Burger (eds), 1985. *Oscillations and traveling waves in chemical systems*, Wiley, New York.

D. Frank-Kamenetskii, 1969. *Diffusion and heat transfer in chemical kinetics*, Plenum, New York.

S. Fraser and R. Kapral, 1982. Analysis of flow hysteresis by a one-dimensional map, *Phys. Rev.* **A25**, 3223–33.

B. Friedman, 1956. *Principles and techniques of applied mathematics*, Dover, New York.

N. Ganapathisubramanian and K. Showalter, 1983. Critical slowing down in the bistable iodate–arsenic (III) reaction, *J. Phys. Chem.* **87**, 1098–9.

F. Gantmacher, 1959. *The theory of matrices*, 2 vols., Chelsea, New York.

C. Gardiner, 1983, *Handbook of stochastic methods*, Springer, Berlin.

P. Gaspard and G. Nicolis, 1983. What can we learn from homoclinic orbits in chaotic dynamics? *J. Stat. Phys.* **31**, 499–518.

P. Gaspard, 1992a. *r*-adic one-dimensional maps and the Euler summation formula, *J. Phys*, **A25**, L483–5.

P. Gaspard, 1992b. Diffusion, effusion and chaotic scattering : an exactly soluble Liouvillian dynamics, *J. Stat. Phys.* **68**, 673–747.

G. Gause, 1934. *The struggle for existence*, Williams and Wilkins, Baltimore.

P. Geysermans and G. Nicolis, 1993. Thermodynamic fluctuations and chemical chaos in a well-stirred reactor : a master equation analysis, *J. Chem. Phys.* **99**, 8964 – 9.

J.W. Gibbs, 1902. *Elementary principles in statistical mechanics*, reprinted in 1960 by Dover, New York.

A. Gierer and H. Meinhardt, 1972. A theory of biological pattern formation, *Kybernetik* **12**, 30–9.

L. Glass and M. Mackey, 1988. *From clocks to chaos: the rhythms of life*, Princeton University Press, Princeton.

P. Glendinning and C. Sparrow, 1984. Local and global behavior near homoclinic orbits, *J. Stat. Phys.* **35**, 645–96.

A. Goldbeter, 1990. *Rythmes et chaos dans les systèmes biochimiques et cellulaires*, Masson, Paris.

H. Goldstein, 1959. *Classical mechanics*, Addison-Wesley, Reading, Mass.

P. Grassberger, R. Badii and A. Politi, 1988. Scaling laws for invariant measures on hyperbolic and nonhyperbolic attractors, *J. Stat. Phys.*, **51**, 135–78.

P. Gray, 1990. Modeling oscillatory behaviour in closed systems, in *Spatial inhomogeneities and transient behaviour in chemical kinetics*, P. Gray, G. Nicolis, F. Baras, P. Borckmans and S. Scott, eds., Manchester University Press, Manchester.

P. Gray, G. Nicolis, F. Baras, P. Borckmans and S. Scott (eds), 1990. *Spatial inhomogeneities and transient behaviour in chemical kinetics*, Manchester University Press, Manchester.

S. Grossmann and S. Thomae, 1977. Invariant distributions and stationary correlation functions of one-dimensional discrete processes, *Z. Naturf.* **32A**, 1353–63.

E. Guazzelli and E. Guyon, 1982. Cusp shaped hydrodynamic instability in a nematic, *J. Phys*, **43**, 985–9.

J. Guckenheimer and Ph. Holmes, 1983. *Nonlinear oscillations, dynamical systems, and bifurcations of vector fields*, Springer, Berlin.

E. Guyon, J.-P. Hulin and L. Petit, 1991. *Hydrodynamique physique*, Inter Editions, Paris.

J. Hale and H. Koçak, 1991. *Dynamics and bifurcations*, Springer, Berlin.

T. Halsey, M. Jensen, L. Kadanoff, I. Procaccia and B. Shraiman, 1986. Fractal measures and their singularities: the characterization of strange sets, *Phys. Rev.* **A33**, 1141–51.

B.L. Hao, 1989. *Elementary symbolic dynamics and chaos in dissipative systems*, World Scientific, Singapore.

H. Hasegawa and W. Saphir, 1992. Unitarity and irreversibility in chaotic systems, *Phys. Rev.* **A46**, 7401–23.

H. Hentschel and I. Procaccia, 1983. The infinite number of generalized dimensions of fractals and strange attractors, *Physica* **8D**, 435–44.

B. Hu and L. Rudnick, 1982. Exact solutions to the renormalization-group fixed-point equations for intermittency in two-dimensional maps, *Phys. Rev.* **A26**, 3035–6.

R. Imbihl, 1992. Spatiotemporal pattern formation in catalytic reactions on single crystal surfaces, *Physica* **A188**, 34–46.

E. Infeld and G. Rowlands, 1990. *Nonlinear waves, solitons and chaos*, Cambridge University Press, Cambridge.

G. Iooss and D. Joseph, 1980. *Elementary stability and bifurcation theory*, Springer, Berlin.

F. Jacob and J. Monod, 1961. Genetic regulatory mechanisms in the synthesis of proteins, *J. Mol. Biol.* **3**, 318–56.

D. Jordan and P. Smith, 1977. *Nonlinear ordinary differential equations*, Clarendon Press, Oxford.

M. Kac, 1959. *Probability and related topics in physical sciences*, Interscience, New York.

K. Kaneko, 1989. Pattern dynamics in spatio-temporal chaos, *Physica* **D34**, 1–41.

L. Kantorovitch and V. Krylov, 1964. *Approximate methods of higher analysis*, Noordhoff, Groningen.

R. Kapral and S. Fraser, 1984. Bistable oscillating states in dissipative dynamical systems : scaling properties and one-dimensional maps. *J. Phys. Chem.* **88**, 4845–52.

M. Kaufman and R. Thomas, 1987. Model analysis of the bases of multistationarity in the humoral immune response, *J. Theor. Biol.* **129**, 141–62.

J. Keener, 1976. Secondary bifurcation in nonlinear diffusion – reaction equations, *Stud. Appl. Math.* **55**, 187–211.

E. Kerner, 1957. A statistical mechanics of interacting biological species, *Bull. Math. Biophys.* **19**, 121–46.

J. Keworkian and J. Cole, 1981. *Perturbation methods in applied mathematics*, Springer, Berlin.

A. Khinchine, 1959. *Mathematical foundations of information theory*, Dover, New York.

V. Kondratiev and E. Nikitin, 1981. *Gas-phase reactions*, Springer, Berlin.

E. Koschmieder, 1981. Experimental aspects of hydrodynamic instabilities, in *Order and fluctuations in equilibrium and nonequilibrium statistical mechanics*, G. Nicolis, G. Dewel and J.W. Turner eds., Wiley, New York.

N. Krylov, 1979. *Works on the foundations of statistical physics*, Princeton University Press, Princeton.

Y. Kuramoto, 1984. *Chemical oscillations, waves and turbulence*, Springer, Berlin.

Y. Kuramoto, 1990. Extended phase dynamics approach to pattern evolution, in *Spatial inhomogeneities and transient behaviour in chemical kinetics*, P. Gray, G. Nicolis, F. Baras, P. Borckmans and S. Scott eds., Manchester University Press, Manchester.

L. Landau and E. Lifshitz, 1959a. *Fluid mechanics*, Pergamon, Oxford.

L. Landau and E. Lifshitz, 1959b. *Statistical physics*, Pergamon, Oxford.

O. Lanford, 1982. A computer-assisted proof of the Feigenbaum conjecture, *Bull. Amer. Math. Soc.* **6**, 427–34.

J. Langer, 1980. Instabilities and pattern formation in crystal growth, *Rev. Mod. Phys.* **52**, 1–28.

J. Langer and H. Müller-Krumbhaar, 1983. Mode selection in a dendritelike nonlinear system, *Phys. Rev.* **A27**, 499–514.

W. Langford, 1979. Periodic and steady mode interactions lead to tori, *SIAM J. Appl. Math.* **37**, 22–48.

J. Laskar and P. Robutel, 1993. The chaotic obliquity of the planets, *Nature* **361**, 608–12.

A. Lasota and M. Mackey, 1985. *Probabilistic properties of deterministic systems*, Cambridge University Press, Cambridge.

R. Lefever and G. Nicolis, 1971. Chemical instabilities and sustained oscillations, *J. Theor. Biol.* **30**, 267–84.

R. Lefever, G. Nicolis and P. Borckmans, 1988. The Brusselator: it does oscillate all the same, *J. Chem. Soc., Faraday Trans. 1*, **84**, 1013–23.

J. Lega, 1989. *Défauts topologiques associés à la brisure de l'invariance de translation dans le temps.* Ph.D. dissertation, University of Nice.

R.S. Li and G. Nicolis, 1981. Bifurcation phenomena in nonideal systems. 2. Effect of nonideality correction on symmetry-breaking transitions, *J. Phys. Chem.* **85**, 1912–18.

R.S. Li, G. Nicolis and H. Frisch, 1981. Bifurcation phenomena in nonideal systems. 1. Effect of the nonideality correction on multiple steady-state transitions, *J. Phys. Chem.* **85**, 1907–12.

A. Libchaber and J. Maurer, 1980. Une expérience de Rayleigh–Bénard de géometrie réduite: multiplication, accrochage et démultiplication de fréquences, *J. Phys.* **41**, Coll. C3–51

A. Lichtenberg and M. Lieberman, 1983. *Regular and stochastic motion*, Springer, Berlin.

C. Lin, 1955. *The theory of hydrodynamic stability*, Cambridge University Press, Cambridge.

P. Linsay, 1981. Period doubling and chaotic behavior in a driven anharmonic oscillator, *Phys. Rev. Lett.* **47**, 1349–52.

E. Lorenz, 1963. Deterministic non – periodic flow, *J. Atmos. Sci.* **20**, 130–41.

A. Lotka, 1924. *Elements of mathematical biology*, Johns Hopkins, Baltimore; reprinted in 1956 by Dover Publ., New York.

M. Lowe and G. Gollub, 1985. Pattern selection near the onset of convection: the Eckhaus instability, *Phys. Rev. Lett.* **55**, 2575–8.

M. Lücke and F. Schank, 1985. Response to parametric modulation near an instability, *Phys. Rev. Lett.* **54**, 1465–8.

L. Lugiato and R. Lefever, 1987. Spatial dissipative structures in passive optical systems, *Phys. Rev. Lett.* **58**, 2209–12.

L. Lugiato, 1992. Spatio-temporal structures. Part I, in *Quantum optics*, P. Mandel ed., *Phys. Reports* **219** nos 3–6, 293–310.

S. Ma, 1976. *Modern theory of critical phenomena*, Benjamin, Reading, Mass.

J. Mahar and B. Matkowsky, 1977. A model biochemical reaction exhibiting secondary bifurcation, *SIAM J. Appl. Math.* **32**, 394–404.

H. Malchow and L. Schimansky-Geier, 1985. *Noise and diffusion in bistable nonequilibrium systems*, Teubner, Leipzig.

B. Mandelbrot, 1977. *Fractals: form, chance, dimension*, Freeman, San Francisco.

P. Manneville and Y. Pomeau, 1980. Different ways to turbulence in dissipative systems, *Physica* **1D**, 219–26.

P. Manneville, 1985. Lyapunov exponents for the Kuramoto–Shivashinsky model, in *Macroscopic modeling of turbulent flows*, U. Frisch *et al.* eds., Springer, Berlin.

P. Manneville, 1991. *Structures dissipatives et turbulence faible*, Alléa, Saclay.

J. Marsden and M. McCracken, 1976. *The Hopf bifurcation and its applications*, Springer, Berlin.

G. Martin, 1983. Long-range periodic decomposition of irradiated solid solutions, *Phys. Rev. Lett.* **50**, 250–2.

R. May, 1974. *Stability and complexity in model ecosystems*, Princeton University Press, Princeton.

R. May, 1976. Simple mathematical models with very complicated dynamics, *Nature* **261**, 459–67.

D. McKernan and G. Nicolis, 1994. Generalized Markov coarse graining and spectral decomposition of chaotic piecewise linear maps, *Phys. Rev.* **E50**, 988–99.

M. Misiurewicz, 1976. Topological conditional entropy, *Studia Mathematica* **55**, 175–200.

E. Montroll and W. Badger, 1974. *Introduction to quantitative aspects of social phenomena*, Gordon and Breach, New York.

J. Moser, 1973. *Stable and random motions in dynamical systems*. Princeton University Press, Princeton.

J. Murray, 1989. *Mathematical biology*, Springer, Berlin.

V. Nemytskii and V. Stepanov, 1960. *Qualitative theory of differential equations*, Princeton University Press, Princeton.

J. Nese, 1989. Quantifying local predictability in phase space, *Physica* **35D**, 237–50.

A. Newell, 1974. Envelope equations, in *Lecture Notes in Appl. Math.* **15**, 157–61, Springer, Berlin.

A. Newell and J. Maloney, 1992. *Nonlinear optics*, Addison-Wesley, Redwood City.

B. Nicolaenko, B. Scheurer and R. Temam, 1985. Some global dynamical properties of the Kuramoto–Shivashinsky equations: nonlinear stability and attractors, *Physica* **D16**, 155–83.

C. Nicolis, 1984. Self-oscillations and predictability in climate dynamics, *Tellus* **36A**, 1–10.

C. Nicolis and G. Nicolis, 1991. Dynamics of error growth in unstable systems, *Phys. Rev.* **A43**, 5720–3.

C. Nicolis, G. Nicolis and Q. Wang, 1992. Sensitivity to initial conditions in spatially distributed systems, *Bifurcation and Chaos* **2**, 263–9.

C. Nicolis and G. Nicolis, 1993. Finite time behavior of small errors in deterministic chaos and Lyapunov exponents, *Bifurcation and Chaos* **3**, 1339–42.

G. Nicolis and J. Portnow, 1973. Chemical oscillations, *Chem. Rev.* **73**, 365–84.

G. Nicolis and J. Auchmuty, 1974. Dissipative structures, catastrophes, and pattern formation : a bifurcation analysis, *Proc. Nat. Acad. Sci. USA* **71**, 2748–51.

G. Nicolis and R. Lefever, 1977. Comment on the kinetic potential and the Maxwell construction in non-equilibrium chemical phase transitions, *Phys. Lett.* **62A**, 469–71.

G. Nicolis and I. Prigogine, 1977. *Self-organization in nonequilibrium systems*, Wiley, New York.

G. Nicolis and J.W. Turner, 1977. Stochastic analysis of a nonequilibrium phase transition : some exact results, *Physica* **89A**, 326–38.

G. Nicolis and F. Baras (eds), 1984. *Chemical instabilities*, Reidel, Dordrecht.

G. Nicolis and C. Nicolis, 1988. Master-equation approach to deterministic chaos, *Phys. Rev.* **A38**, 427–33.

G. Nicolis, C. Nicolis and J.S. Nicolis, 1989. Chaotic dynamics, Markov partitions and Zipf's law, *J. Stat. Phys.* **54**, 915–24.

G. Nicolis, Y. Piasecki and D. McKernan, 1992. Toward a probabilistic description of deterministic chaos, in *From phase transitions to chaos*, G. Györgyi, I. Kondor, L. Sasvári and T. Tél eds., World Scientific, Singapore.

J.S. Nicolis, G. Mayer-Kress and G. Haubs, 1983. Non-uniform chaotic dynamics with implications to information processing. *Z. Naturf.* **A38**, 1157–69.

J.S. Nicolis, 1991. *Chaos and information processing*, World Scientific, Singapore.

R. Noyes and R. Field, 1974. Oscillatory chemical reactions, *Ann. Rev. Phys. Chem.* **25**, 95–119.

P. Ortoleva and J. Ross, 1973. Phase waves in oscillatory chemical reactions, *J. Chem. Phys.* **58**, 5673–80.

H. Othmer, 1976. Nonuniqueness of equilibria in closed reacting systems, *Chem. Eng. Sci.* **31**, 993–1003.

Q. Ouyang and H. Swinney, 1991. Transition from uniform state to hexagonal and striped Turing patterns, *Nature* **352**, 610–12.

A. Pacault and C. Vidal (eds) 1978. *Far from equilibrium*, Springer, Berlin.

E. Patterson, 1959. *Topology*, Oliver and Boyd, Edinburgh.

O. Penrose, 1970. *Foundations of statistical mechanics*, Pergamon, Oxford.

O. Penrose, 1979. Foundations of statistical mechanics, *Rep. Prog. Phys.* **42**, 1937–2006.

J. Pesin, 1977. Characteristic Lyapunov exponents and smooth ergodic theory, *Russ. Math. Surv.* **32**, 55–114.

A. Pocheau, V. Croquette and P. Le Gal, 1985. Turbulence in a cylindrical container of argon near threshold of convection, *Phys. Rev. Lett.* **55**, 1094–7.

Y. Pomeau, A. Pumir and P. Pelcé, 1984. Intrinsic stochasticity with many degrees of freedom, *J. Stat. Phys.* **37**, 39–49.

I. Prigogine, 1947. *Etude thermodynamique des processus irréversibles*, Desoer, Liége.

I. Prigogine, 1962. *Nonequilibrium statistical mechanics*, Wiley-Interscience, New York.

I. Prigogine and R. Lefever, 1968. Symmetry-breaking instabilities in dissipative systems, *J. Chem. Phys.* **48**, 1695–700.

I. Prigogine, 1980. *From being to becoming*, Freeman, San Francisco.

R. Rand and D. Armbruster, 1987. *Perturbation methods, bifurcation theory and computer algebra*, Springer, Berlin.

A. Renyi, 1970. *Probability theory*, North Holland, Amsterdam.

P. Richetti, 1987. *Etude théorique et numérique des dynamiques chaotiques de la réaction de Belousov–Zhabotinski*, Ph.D. dissertation, University of Bordeaux.

P. Richetti and A. Arnéodo, 1985. The periodic-chaotic sequences in chemical reactions : a scenario close to homoclinic conditions?, *Phys. Lett.* **A109**, 359–66.

O. Rössler, 1976. An equation for continuous chaos, *Phys. Lett.* **57A**, 397–8.

O. Rössler, 1979. Continuous chaos – four prototype equations, *Ann. N.Y. Acad. Sci.* **316**, 376–92.

D. Ruelle and F. Takens, 1971. On the nature of turbulence, *Commun. Math. Phys.* **20**, 167–92.

D. Ruelle, 1985. Resonances of chaotic dynamical systems, *Phys. Rev. Lett.* **56**, 405–7.

D. Ruelle, 1986. Locating resonances for Axiom A dynamical systems, *J. Stat. Phys.* **44**, 281–92.

B. Saltzman, A. Sutera and A. Hansen, 1982. A possible marine mechanism for internally generated long-period climate cycles, *J. Atmos. Sci.* **39**, 2634–7.

D. Sattinger, 1972. *Topics in stability and bifurcation theory*, Springer, Berlin.

F. Schlögl, 1971. On thermodynamics near a steady state, *Z. Phys.* **248**, 446–58.

F. Schlögl, 1972. Chemical reaction models for nonequilibrium phase transitions, *Z. Phys.* **253**, 147–61.

M. Schröder, 1991. *Fractals, chaos and power laws*, Freeman, New York.

H. G. Schuster, 1988. *Deterministic chaos*, VCH Verlag, Weinheim.

L. Segel, 1984. *Modeling dynamical phenomena in molecular and cellular biology*, Cambridge University Press, Cambridge.

C. Shannon and W. Weaver, 1949. *The mathematical theory of communication*, University of Illinois Press, Urbana.

L. Shil'nikov, 1965. A case of existence of a denumerable set of periodic motions, *Sov. Math. Dokl.* **6**, 163–6.

R. Simoyi, A. Wolf and H. Swinney, 1982. One-dimensional dynamics in a multicomponent chemical reaction, *Phys. Rev. Lett.* **49**, 245–8.

I. Sokolnikoff, 1939. *Advanced calculus*, McGraw-Hill, New York.

C. Sparrow, 1982. *The Lorenz equations*, Springer, Berlin.

H.E. Stanley, 1971. *Introduction to phase transitions and critical phenomena*, Clarendon, Oxford.

G. Sussman and J. Wisdom, 1992. Chaotic evolution of the solar system, *Science* **257**, 56–62.

T. Tel, 1987. Fractals and multifractals, Technical report, Eötvos University, Budapest.

R. Thom, 1962. *Stabilité structurelle et morphogénèse*, Benjamin, New York.

R. Thomas and R. D'Ari, 1990. *Biological feedback*, CRC press, Boca Raton, Florida.

M. Thompson, 1982. *Instabilities and catastrophes in science and engineering*, Wiley, Chichester.

A. Turing, 1952. The chemical basis of morphogenesis, *Phil. Trans. Roy. Soc. London* **B237**, 37–72.

J.S. Turner, J.C. Roux, W. McCornick and H. Swinney, 1981. Alternating periodic and chaotic regimes in a chemical reaction-experiment and theory, *Phys. Lett.* **85A**, 9–12.

J. Tyson, 1976. *The Belousov–Zhabotinski reaction*, Springer, Berlin.

S. Vannitsem and C. Nicolis, 1994. Predictability experiments on a simplified thermal convection model : the role of spatial scales, *J. Geophys. Res.* **99**, 10377–85.

M. Velarde and R. Perez Gordon, 1976. On the (nonlinear) foundations of Boussinesq approximation applicable to a thin layer of fluid. II, *J. Phys.* **37**, 177–82.

M. Velarde and C. Normant, 1980. Convection, *Scientific American* **243**(1), 78–108.

P.F. Verhulst, 1845. Recherches mathématiques sur la loi d'accroissement de la population, *Mem. Acad. Roy. Sci. Belles Lett. Brux.* **18**, 1–38.

C. Vidal and A. Pacault (eds.), 1981. *Nonlinear phenomena in chemical dynamics*, Springer, Berlin.

C. Vidal and A. Pacault (eds.), 1984. *Nonequilibrium dynamics in chemical systems*, Springer, Berlin.

V. Volterra, 1936. *Leçons sur la théorie mathématique de la lutte pour la vie*, Gauthier-Villars, Paris.

D. Walgraef, G. Dewel and P. Borckmans, 1982. Nonequilibrium phase transitions and chemical instabilities, *Adv. Chem. Phys.* **49**, 311–55.

D. Walgraef, 1988. *Structures spatiales loin de l'équilibre*, Masson, Paris.

X.J. Wang, 1983. *Bifurcations de Hopf en interaction et comportements quasi-périodiques dans les systèmes dissipatifs non-linéaires*, Mémoire de licence, University of Brussels.

X.J. Wang and G. Nicolis, 1987. Bifurcation phenomena in coupled chemical oscillators : normal form analysis and numerical simulations, *Physica* **26D**, 140–55.

W. Wasow, 1965. *Asymptotic expansions for ordinary differential equations*, Interscience, New York.

S. Wiggins, 1990. *Introduction to applied nonlinear dynamical systems and chaos*, Springer, Berlin.

K. Willamowski and O. Rössler, 1980. Irregular oscillations in a realistic abstract quadratic mass action system, *Z. Naturf.* **35a**, 317–18.

L. Wolpert, 1969. Positional information and the spatial pattern of cellular differentiation, *J. Theor. Biol.* **25**, 1–47.

Y. Zeldovich, G. Barenblatt, V. Librovich and G. Makhviladze, 1985. *Mathematical theory of combustion and explosions*, Plenum, New York.

A. Zhabotinski, 1964. Periodic oxidation reactions in the liquid phase, *Dokl. Akad. Nauk. SSSR* (English transl) **157**, 701–4.

Index

Printed in the United States
By Bookmasters